SRA
Connecting Math Concepts

Level C Textbook

COMPREHENSIVE EDITION

A DIRECT INSTRUCTION PROGRAM

Education

Bothell, WA • Chicago, IL • Columbus, OH • New York, NY

MHEonline.com

 **Education**

Send all inquiries to:
McGraw-Hill Education
4400 Easton Commons
Columbus, OH 43219

ISBN: 978-0-02-103578-6
MHID: 0-02-103578-4

Printed in the United States of America.

6 7 8 9 10 QVS 18 17 16 15 14

The McGraw-Hill Companies

Lesson

Part 1

a. Jane started out with 56 boxes.
    She made 20 more boxes.
    How many boxes did she end up with?

b. Cara had 43 glasses.
    Then she broke 21 glasses.
    How many glasses did she end up with?

c. Donna had some stickers.
    She bought 13 more stickers.
    She ended up with 36 stickers.
    How many did she start out with?

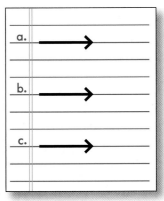

Part 2

a. $10 \times 6 =$

b. $9 \times 4 =$

c. $2 \times 5 =$

d. $4 \times 2 =$

Lesson

a. A bus had 13 people on it.
Then 16 more people got on the bus.
How many people ended up on the bus?

b. A train had some people on it.
Then 19 people got on the train.
The train ended up with 89 people on it.
How many people did the train start with?

c. A train had 97 people on it.
Then 34 people got off the train.
How many people ended up on the train?

d. A train had some people on it.
Then 155 people got off the train.
The train ended up with 23 people.
How many people did the train start with?

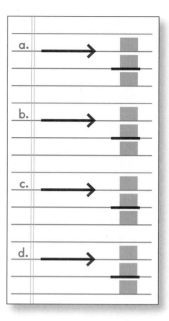

Part 2

a.	5 x 9 =
b.	9 x 5 =
c.	2 x 6 =

Lesson 43

a. A truck had chickens on it.
Then the driver put 33 more chickens on the truck.
The truck ended up with 55 chickens on it.
How many chickens did the truck start out with?

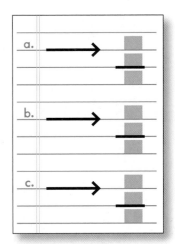

b. Millie had 40 dollars.
She earned 25 dollars.
How many dollars did Millie end up with?

c. At the start of the day, Rob had 68 marbles.
Then he gave away 26 marbles.
How many marbles did Rob end up with?

Independent Work

Part 2 Copy each problem and work it.

a. 5 x 7 =

b. 1 x 6 =

c. 4 x 2 =

d. 10 x 2 =

Lesson

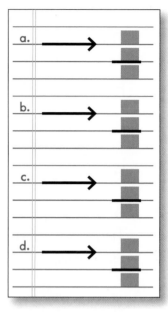

a. A train started out with 60 people.
   Then 31 more people got on the train.
   How many people ended up on the train?

b. A train started out with some people.
   Then 81 more people got on the train.
   The train ended up with 194 people.
   How many people started out on the train?

c. A train started out with some people on it.
   Then 12 people got off the train.
   The train ended up with 100 people.
   How many people started out on the train?

d. A train started out with 90 people.
   Then 40 people got off the train.
   How many people ended up on the train?

Independent Work

Part 2 Copy each problem and work it.

a. $5 \times 4 =$

b. $4 \times 5 =$

c. $9 \times 6 =$

Lesson 45

Part 1

a. A truck started out with some packages.
   It dropped off 15 packages.
   It ended up with 21 packages.
   How many packages did the truck start out with?

b. Mark started out with 28 books.
   He gave away 6 books.
   How many books did Mark end up with?

c. Rob had 12 shells at home.
   Then he found 14 more shells on the beach.
   How many shells did Rob end up with?

d. Faith had some shoes in a box.
   She found 10 shoes.
   She ended up with 18 shoes.
   How many shoes did Faith start with?

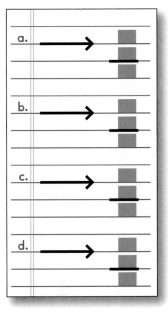

Part 2

a.

b.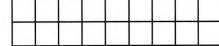

a.

b.

Independent Work

Part 3   Copy each problem and work it.

a. $5 \times 3 =$	d. $4 \times 5 =$
b. $9 \times 4 =$	e. $5 \times 4 =$
c. $10 \times 4 =$	f. $2 \times 6 =$

Lesson 46

Part 1

a. Tim had 36 dimes.
   Then he earned 11 more dimes.
   How many dimes did Tim end up with?

b. Jim had some toy cars.
   He bought 16 new cars.
   He ended up with 29 cars.
   How many cars did Jim start with?

c. Sam had 154 buttons.
   Then he gave away 53 buttons.
   How many buttons did Sam end up with?

d. A dog started out with some fleas.
   Then the dog got rid of 40 fleas.
   The dog ended up with 25 fleas.
   How many fleas did the dog start out with?

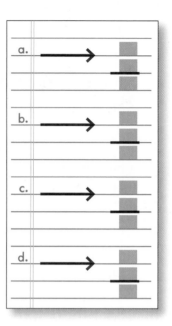

Lesson 46

Part 2

a.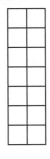

b.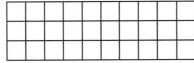

c.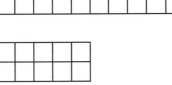

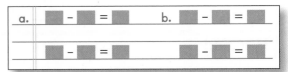

a. ▨ x ▨ = ▨

b. ▨ x ▨ = ▨

c. ▨ x ▨ = ▨

Part 3 Write 2 subtraction facts for each family.

a.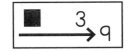
■ ──3──→ 9

b.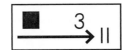
■ ──3──→ 11

a. ▨ – ▨ = ▨   b. ▨ – ▨ = ▨

▨ – ▨ = ▨   ▨ – ▨ = ▨

Independent Work

Part 4 Copy each problem and work it.

a. 2 x 7 =   d. 4 x 1 =

b. 9 x 2 =   e. 4 x 3 =

c. 1 x 4 =   f. 5 x 2 =

Part 1

a.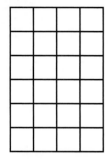

b.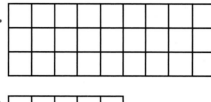

c.

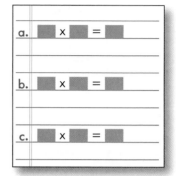

a. ☐ x ☐ = ☐

b. ☐ x ☐ = ☐

c. ☐ x ☐ = ☐

Independent Work

Part 2

a. Ernie had 16 boxes.
   He made 13 more boxes.
   How many boxes did he end up with?

b. Mrs. Johnson had some cakes.
   She gave away 9 cakes.
   She ended up with 7 cakes.
   How many cakes did she start with?

c. The children had bags of popcorn.
   They filled 12 more bags of popcorn.
   They ended up with 18 bags.
   How many bags of popcorn did they start with?

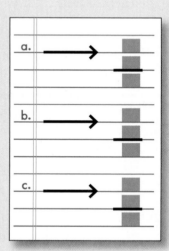

Part 3   Copy and work each problem.

a. $2 \times 10 =$	d. $4 \times 5 =$	
b. $5 \times 8 =$	e. $10 \times 2 =$	
c. $9 \times 3 =$	f. $1 \times 6 =$	

Lesson

Part 1

a.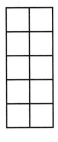

b.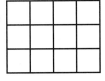

c.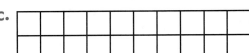

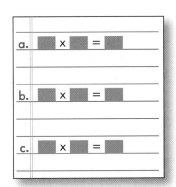

a. ☐ x ☐ = ☐

b. ☐ x ☐ = ☐

c. ☐ x ☐ = ☐

Part 2

a. Heidi has 17 more marbles than Bill has.
Heidi has 48 marbles.
How many marbles does Bill have?

b. Sarah made 10 more cupcakes than Maria made.
Maria made 24 cupcakes.
How many cupcakes did Sarah make?

c. Hank's car is 8 years older than Tim's car.
Hank's car is 11 years old.
How old is Tim's car?

d. Bob has 7 dollars less than Val has.
Bob has 31 dollars.
How many dollars does Val have?

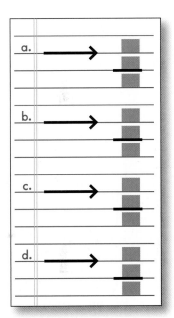

Independent Work

Part 3   Write 2 subtraction facts for each family.

a.

b.

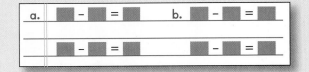

a. ☐ – ☐ = ☐   b. ☐ – ☐ = ☐

☐ – ☐ = ☐   ☐ – ☐ = ☐

Lesson 48

Part 4 Copy and work each problem.

a. 9 x 5 =	d. 2 x 4 =
b. 5 x 9 =	e. 4 x 2 =
c. 9 x 2 =	f. 5 x 2 =

Part 5

a. Karen had some dollars.
   She spent 44 dollars.
   She ended up with 40 dollars.
   How many dollars did she start with?

b. Frank had 24 pretzels.
   His dog ate 20 of those pretzels.
   How many pretzels did Frank end up with?

c. Bill had 40 goldfish.
   Then he bought 15 goldfish.
   How many goldfish did Bill end up with?

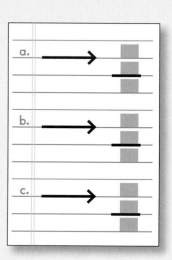

Lesson 49

Part 1

a.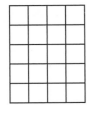

b.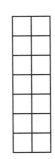

c.

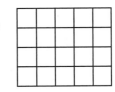

Part 2

a. Willy has 9 more teeth than Dilly has.
Dilly has 3 teeth.
How many teeth does Willy have?

b. Heidi weighed 21 pounds less than Bob.
Bob weighed 145 pounds.
How many pounds did Heidi weigh?

c. Fran bought 24 more bags of corn than Ann bought.
Ann bought 11 bags of corn.
How many bags did Fran buy?

d. Tina is 13 years younger than Ron.
Tina is 16.
How old is Ron?

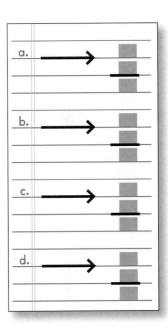

Independent Work

Part 3   Copy and work each problem.

a. 5 x 9 =   d. 2 x 5 =

b. 9 x 5 =   e. 4 x 3 =

c. 10 x 3 =   f. 9 x 3 =

Lesson 49

Part 4

a. Sam had 34 seashells.
Then he lost 11 seashells.
How many seashells did he end up with?

b. Maria had some buckets in her store.
She sold 12 buckets.
She ended up with 14 buckets.
How many buckets did she start with?

c. First, Rod wrote 18 songs.
Then he wrote 13 more songs.
How many songs did he end up writing?

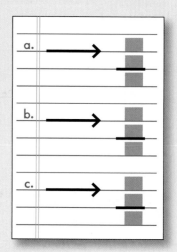

Lesson 50

Part 1

a. Ginger had 14 dogs.
   Then she sold some dogs.
   She ended up with 11 dogs.
   How many dogs did she sell?

b. A train started out with 90 people.
   Then some people got off the train.
   The train ended up with 10 people.
   How many people got off the train?

c. A train started out with 46 people.
   Some more people got on the train.
   The train ended up with 58 people on it.
   How many more people got on the train?

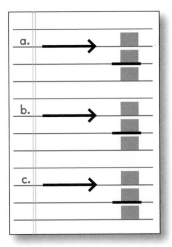

Part 2

a. Sarah made 6 fewer pies than Maria made.
   Sara made 20 pies.
   How many pies did Maria make?

b. Jack had 2 fewer teeth than Mary had.
   Mary had 26 teeth.
   How many teeth did Jack have?

c. Sid bought 8 more things than Gail bought.
   Gail bought 1 thing.
   How many things did Sid buy?

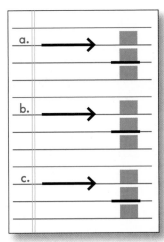

Lesson 50

Part 3 Write the letter **C**, **T**, or **R**. Then write **S** for each square.

a. 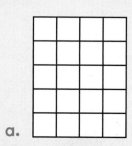 b. c. d. e.

a.	
b.	
c.	
d.	
e.	

Part 4 Write the problem for **a**. Then copy and work problems **b–f**.

a.

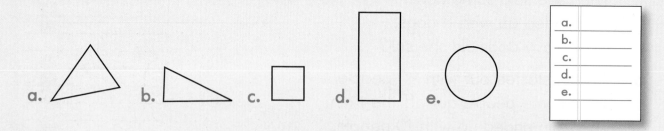

a. ■ x ■ = ■   d. 2 x 7 = ■

b. 9 x 4 = ■   e. 5 x 6 = ■

c. 1 x 5 = ■   f. 10 x 3 = ■

Part 5

a. K is 26 more than M.
    M is 50.
    What number is K?

b. R is 32 less than P.
    R is 61.
    What number is P?

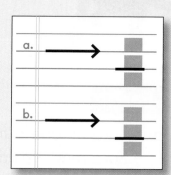

Lesson 51

Part 1

a. A girl had 13 books.
    Then her sister gave her more books.
    She ended up with 23 books.
    How many books did her sister give her?

b. A girl had some books.
    Her sister gave her 22 books.
    She ended up with 45 books.
    How many books did she start with?

c. A girl had 9 books.
    She gave away some books.
    She ended up with 2 books.
    How many books did she give away?

d. A girl had 18 books.
    Then she bought 11 books.
    How many books did she end up with?

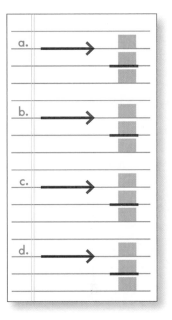

Lesson 51

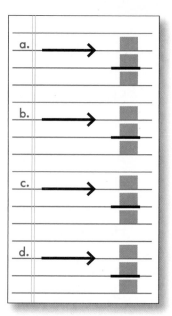

Part 2

a. The cat was 9 pounds lighter than the dog.
The cat weighed 11 pounds.
How many pounds did the dog weigh?

b. The garage was 12 years older than the house.
The house was 56 years old.
How many years old was the garage?

c. Kim had 11 fewer books than Dan had.
Kim had 33 books.
How many books did Dan have?

d. Scott had 16 fewer tools than Jim had.
Jim had 48 tools.
How many tools did Scott have?

Independent Work

Part 3
Write the problem for **a** and work it. Then copy and
work problems **b–e**.

a.

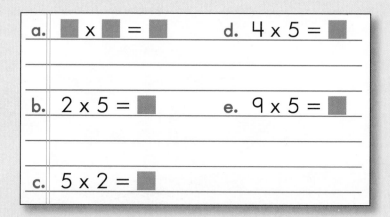

a. $\blacksquare \times \blacksquare = \blacksquare$ d. $4 \times 5 = \blacksquare$

b. $2 \times 5 = \blacksquare$ e. $9 \times 5 = \blacksquare$

c. $5 \times 2 = \blacksquare$

Part 4
Write the letter **C, T,** or **R.** Then write **S** for each square.

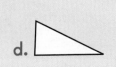

a. 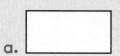   b.   c.   d.

Lesson 52

Part 1

a. Mike had some paper.
He used up 20 sheets of paper.
He ended up with 14 sheets of paper.
How many sheets of paper did Mike start with?

b. Sam started with 6 cards.
He made some more cards.
He ended up with 19 cards.
How many cards did Sam make?

c. Ray had 101 bolts.
He bought 52 more bolts.
How many bolts did Ray end up with?

d. Molly started with 58 peanuts.
She ate some peanuts.
She ended up with 26 peanuts.
How many peanuts did Molly eat?

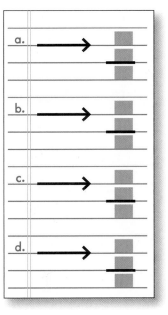

Part 2

a. Rob was 28 years younger than Hilda.
Rob was 41 years old.
How many years old was Hilda?

b. The car was 200 pounds heavier than the log.
The car weighed 953 pounds.
How many pounds did the log weigh?

c. Jan is 13 inches taller than her brother.
Her brother is 44 inches tall.
How many inches tall is Jan?

d. There are 26 fewer sheep than goats.
There are 72 sheep.
How many goats are there?

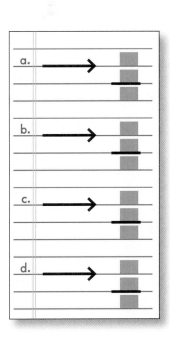

Lesson 52

Part 3 Write 2 subtraction facts.

a.

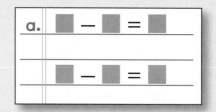

Part 4 Write the problem for **a** and work it. Then copy and work problems **b–d**.

a.

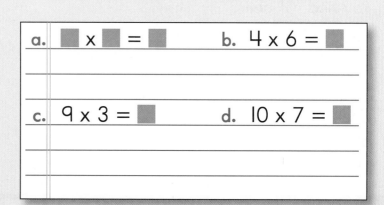

Part 5

a. R is 73 less than V.
R is 106.
What number is V?

b. K is 19 more than T.
T is 80.
What number is K?

c. P is 36 less than F.
F is 49.
What number is P?

d. W is 57 less than Y.
Y is 89.
What number is W?

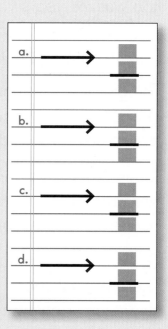

Connecting Math Concepts

Lesson 53

Part 1

a. Anna had 50 feet of ribbon.
She gave away 20 feet of ribbon.
How many feet of ribbon did Anna end up with?

b. Anna started with 12 buttons.
Mona gave her some buttons.
She ended up with 87 buttons.
How many buttons did Mona give her?

c. Joe had some plates.
He broke 6 plates.
He ended up with 12 plates.
How many plates did Joe start with?

d. The farm had 560 cows.
The farm sold some cows.
The farm ended up with 20 cows.
How many cows did the farm sell?

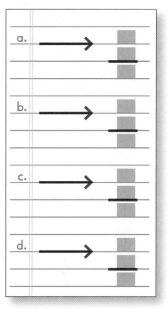

Part 2

a. The <u>w</u>hite dog was 11 pounds heavier than the
<u>b</u>rown dog.
The brown dog weighed 33 pounds.
How many pounds did the white dog weigh?

b. <u>J</u>ane is 5 inches taller than <u>T</u>im.
Jane is 65 inches tall.
How many inches tall is Tim?

c. <u>T</u>im was 9 pounds lighter than <u>B</u>ill.
Tim weighed 90 pounds.
How many pounds did Bill weigh?

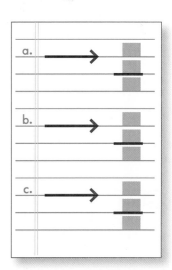

Lesson 53

Part 3 Write the problem for **a** and work it. Then copy and work problems **b–d.**

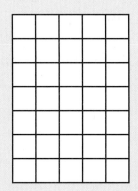

a.

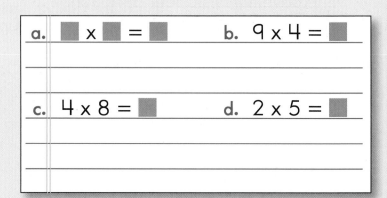

a. ■ x ■ = ■   b. 9 x 4 = ■

c. 4 x 8 = ■   d. 2 x 5 = ■

Part 4

a. M is 51 more than P.
    M is 70.
    What number is P?

b. T is 60 less than K.
    K is 97.
    What number is T?

c. W is 29 more than V.
    W is 81.
    What number is V?

d. N is 18 more than R.
    R is 61.
    What number is N?

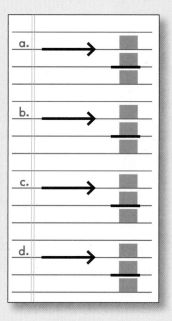

Part 5 Write the letter **C, T,** or **R.** Then write **S** for each square.

a.   b.   c. d.   e.

a.	
b.	
c.	
d.	
e.	

Connecting Math Concepts

Lesson 54

Part 1

a. Mr. Briggs had 16 watermelons.
    Then he bought some more watermelons.
    He ended up with 20 watermelons.
    How many watermelons did he buy?

b. Jane had 32 pencils.
    Then she lost 10 pencils.
    How many pencils did she end up with?

c. Ray's car started out with 17 gallons of gas.
    At the end of the trip, the car had 5 gallons of gas.
    How many gallons did Ray use on the trip?

d. Alex had some money.
    He gave his father 36 dollars.
    He ended up with 20 dollars.
    How many dollars did Alex start with?

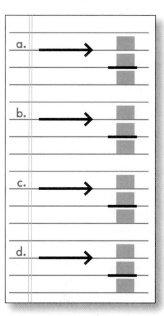

Part 2

a. Sarah has 12 fewer marbles than Julie has.
    Sarah has 16 marbles.
    How many marbles does Julie have?

b. Bob's tent is 2 pounds lighter than Rita's tent.
    Rita's tent weighs 23 pounds.
    How many pounds does Bob's tent weigh?

c. Ron is 13 years older than Mary.
    Mary is 15 years old.
    How many years old is Ron?

d. Ray is 20 pounds heavier than Jill.
    Ray weighs 95 pounds.
    How many pounds does Jill weigh?

e. Joe is 17 years younger than Nick.
    Joe is 62 years old.
    How many years old is Nick?

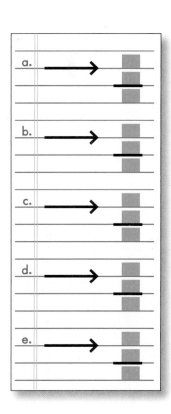

Independent Work

Part 3 Write 2 subtraction facts.

a.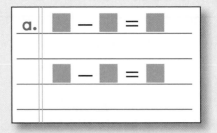

a. ⬛ − ⬛ = ⬛

⬛ − ⬛ = ⬛

Part 4

a. J is 18 less than P.
    J is 41.
    What number is P?

b. T is 39 less than R.
    R is 89.
    What number is T?

c. Y is 42 more than F.
    F is 33.
    What number is Y?

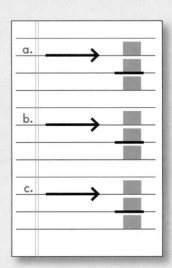

Part 5 Write the problem for **a** and work it. Then copy and work problems **b–e**.

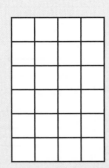

a.

a. ⬛ x ⬛ = ⬛   d. 5 x 5 = ⬛

b. 9 x 3 = ⬛   e. 10 x 6 = ⬛

c. 1 x 10 = ⬛

Lesson 55

Part 1

a. A bus had 34 people on it. Then some more people got on the bus. Now there are 70 people on the bus. How many people got on the bus?

b. Rob had 28 flowers in his garden. 15 more flowers grew. How many flowers ended up in the garden?

c. Bill started out with some trees. He chopped down 13 trees. He ended up with 55 trees. How many trees did he start with?

d. 45 people were at a party. After some people left, there were only 20 people at the party. How many people left the party?

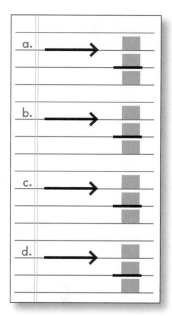

Part 2

Write the letter or letters for each shape.

1.   2.   3.   4.

5.   6.   7.   8.

9.   10. 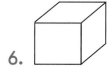   11.   12.

| Pyramid | Cube | Sphere |

1.	2.	3.	4.
5.	6.	7.	8.
9.	10.	11.	12.

Lesson

a. Billy   b. Kelly   c. Hillary   d. Alice

Part 4

a. Alice is 7 years younger than Kelly.
Alice is 31 years old.
How many years old is Kelly?

b. Hillary has 18 more coins than Alice.
Hillary has 59 coins.
How many coins does Alice have?

c. Billy is 15 pounds heavier than Alice.
Alice weighs 119 pounds.
How many pounds does Billy weigh?

d. Kelly has 23 fewer books than Hillary.
Hillary has 48 books.
How many books does Kelly have?

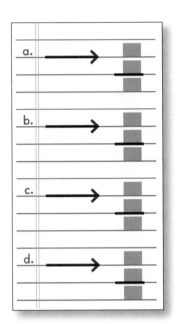

Independent Work

Part 5   Work the problem for **a.** Then copy and work problems **b–g.**

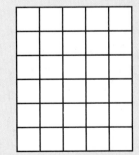

a.

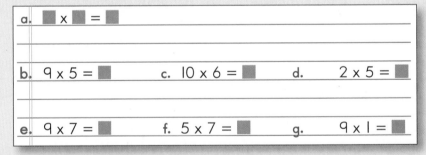

a. ▮ x ▮ = ▮

b. 9 x 5 = ▮   c. 10 x 6 = ▮   d. 2 x 5 = ▮

e. 9 x 7 = ▮   f. 5 x 7 = ▮   g. 9 x 1 = ▮

Part 6

a. R is 15 more than P.
R is 46. What number is P?

b. K is 19 less than T.
T is 71. What number is K?

c. M is 29 more than R.
R is 53. What number is M?

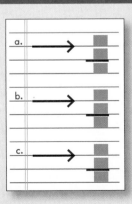

Lesson

a. Rob had some money. He spent 30 dollars and ended up with 140 dollars. How many dollars did he start with?

b. Kris had 89 stickers. She lost some stickers. She ended up with 46 stickers. How many stickers did she lose?

c. Mark had 45 balloons ready for his party. He blew up 22 more balloons. How many balloons did he end up with?

d. Will had 40 nickels. His dad gave him some more nickels. He ended up with 65 nickels. How many nickels did his dad give to Will?

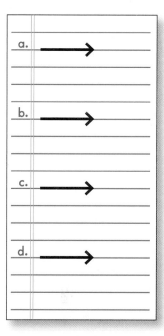

Part 2

a. Sandra has 15 fewer fish than Molly.
Molly has 36 fish.
How many fish does Sandra have?

b. Ryan is 20 years older than Mike.
Ryan is 28.
How many years old is Mike?

c. Paul is 19 pounds lighter than Tim.
Paul weighs 75 pounds.
How many pounds does Tim weigh?

d. Will has 21 fewer dollars than Ray.
Ray has 52 dollars.
How many dollars does Will have?

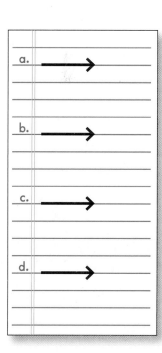

Lesson

Part 3 Write the letter or letters for each shape.

1. [rectangle]   2. [pyramid/triangle]   3. [sphere/egg]

4. [square]   5. [circle]   6. [cube]

Pyramid	Cube	Sphere

1.	2.	3.
4.	5.	6.

Part 4 Write the problem for **a** and work it. Then copy and work problems **b–f.**

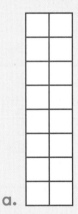

a.

a. ■ x ■ = ■   b. 2 x 7 = ■   c. 10 x 4 = ■

d. 9 x 3 = ■   e. 2 x 9 = ■   f. 5 x 7 = ■

Part 5 Write 2 subtraction facts for each family.

a. ■ —3→ 11   b. ■ —3→ 12   c. ■ —3→ 7

a. ■ — ■ = ■   b. ■ — ■ = ■   c. ■ — ■ = ■

■ — ■ = ■   ■ — ■ = ■   ■ — ■ = ■

Lesson

a. Jan had 15 buttons. She bought some buttons. She ended up with 29 buttons. How many buttons did she buy?

b. Mike had 140 baseball cards. His friend gave him 20 cards. How many cards did he end up with?

c. There were 259 students at the school. 52 students left for a field trip. How many students ended up at the school?

d. Ray had some crayons. He lost 18 crayons. He ended up with 24 crayons. How many crayons did he start with?

a. ⟶

b. ⟶

c. ⟶

d. ⟶

Part 2

a. Truck M was 14 feet shorter than truck R.
Truck M was 35 feet long.
How many feet long was truck R?

b. Fran was 7 inches taller than Jan.
Jan was 51 inches tall.
How many inches tall was Fran?

c. Jim was 3 inches shorter than Hank.
Hank was 74 inches tall.
How many inches tall was Jim?

d. The bike was 17 pounds heavier than the chair.
The bike weighed 29 pounds.
How many pounds was the chair?

a. ⟶

b. ⟶

c. ⟶

d. ⟶

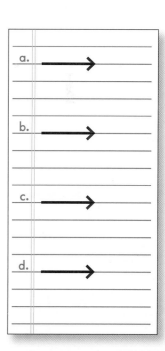

Lesson

Part 3 Write the letter or letters for each shape.

1.   2. ▢ 3. ▢

Pyramid	Cube	Sphere

1.	2.	3.
4.	5.	6.

4. 🔺 5. ⬡ 6. ◹

Part 4 Write the problem for **a** and work it. Then copy and work problems **b–g.**

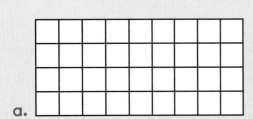

a.

a. ■ x ■ = ■   e. 1 x 5 = ■

b. 2 x 6 = ■   f. 10 x 6 = ■

c. 5 x 8 = ■   g. 9 x 6 = ■

d. 4 x 4 = ■

Part 5

a. P is 12 less than J.
   J is 60. What number is P?

b. T is 19 more than Z.
   Z is 42. What number is T?

c. K is 78 less than Y.
   K is 13. What number is Y?

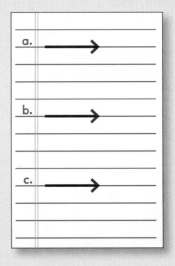

Connecting Math Concepts

Lesson 58

Part 1

a. There were 127 flowers in the garden. Then someone planted 42 flowers. How many flowers are in the garden now?

b. Mandy had a lot of pens in the drawer. Then she put 16 more pens in the drawer. She ended up with 48 pens in the drawer. How many pens did she start with?

c. Rick started with lots of wood. He burned 303 pounds of wood. He ended up with 526 pounds of wood. How many pounds of wood did Rick start with?

d. Dave had 85 dollars. He lost some and ended up with 50 dollars. How many dollars did he lose?

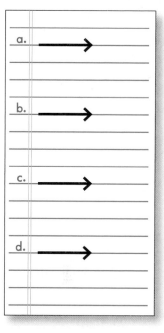

Part 2

a. blue truck
red truck

b. brown tent
yellow tent

c. brick house
wood house

Part 3

a. A red truck is 12 feet shorter than a blue truck.
The red truck is 41 feet long.
How many feet long is the blue truck?

b. The brown tent is 19 years older than the yellow tent.
The yellow tent is 2 years old.
How many years old is the brown tent?

c. The brick house is 10 feet taller than the wood house.
The brick house is 35 feet tall.
How many feet tall is the wood house?

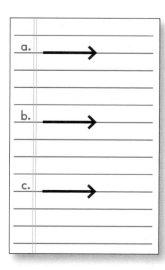

Lesson 58

Independent Work

Part 4 Write the problem for **a** and work it. Then copy and work problems **b–g.**

a.

a. ■ x ■ = ■	e. 10 x 2 = ■
b. 5 x 3 = ■	f. 2 x 9 = ■
c. 9 x 2 = ■	g. 5 x 9 = ■
d. 9 x 5 = ■	

Part 5 Write 2 subtraction facts for each family.

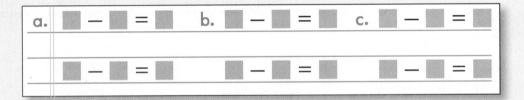

a. [■ →3→ 9]   b. [■ →3→ 12]   c. [■ →3→ 10]

a. ■ – ■ = ■   b. ■ – ■ = ■   c. ■ – ■ = ■

■ – ■ = ■   ■ – ■ = ■   ■ – ■ = ■

Part 6

a. P is 88 less than T.
   T is 199. What number is P?

b. K is 72 less than R.
   K is 125. What number is R?

c. Q is 7 more than X.
   Q is 90. What number is X?

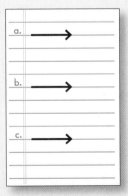

a. →
b. →
c. →

Lesson 59

Part 1

a. Joe started with some nails. He got 36 more nails. He ended up with 79 nails. How many nails did he start with?

b. Bill had 485 stamps. He used 125 stamps. How many stamps did Bill end up with?

c. A store had 280 cans. It got 315 more cans. How many cans did the store end up with?

d. A hotel had 588 glasses. Some glasses broke. The hotel still had 528 glasses. How many glasses broke?

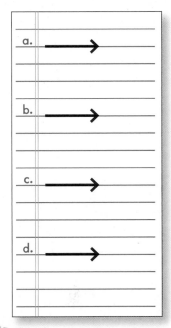

Part 2

a. A red barn is 18 feet taller than a white barn.
The red barn is 38 feet tall.
How many feet tall is the white barn?

b. There are 28 fewer green marbles than yellow marbles.
There are 13 green marbles.
How many yellow marbles are there?

c. The deck is 12 feet shorter than the fence.
The fence is 36 feet long.
How many feet long is the deck?

d. The stone wall is 8 feet longer than the brick wall.
The brick wall is 23 feet long.
How many feet long is the stone wall?

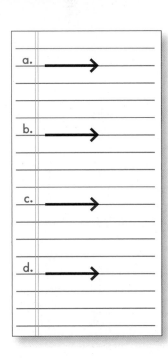

Lesson 59

Part 3

Write the letter or letters for each shape.

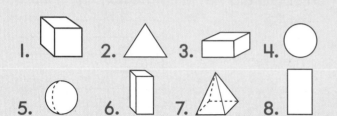

1. 2. 3. 4.

5. 6. 7. 8.

Rectangular Prism			
1.	2.	3.	4.
5.	6.	7.	8.

Part 4

a. N is 21 less than R.
R is 70. What number is N?

b. T is 45 less than J.
J is 75. What number is T?

c. P is 15 more than K.
K is 32. What number is P?

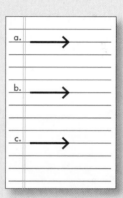

Part 5

Write the problem for **a** and work it. Then copy and work problems **b–g**.

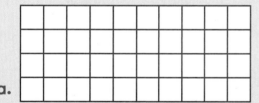

a.

a. ■ x ■ = ■   e. 9 x 2 = ■

b. 5 x 10 = ■   f. 1 x 5 = ■

c. 10 x 5 = ■   g. 5 x 1 = ■

d. 2 x 9 = ■

Part 6

Write 2 subtraction facts for each family.

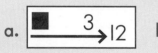

a. ■ —3→ 12

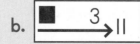

b. ■ —3→ 11

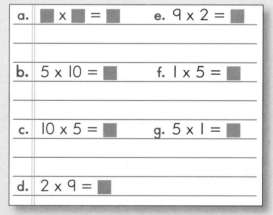

a. ■ — ■ = ■   b. ■ — ■ = ■

■ — ■ = ■   ■ — ■ = ■

Lesson 60

Part 1

a. How much more is 88 than 77?

b. How much less is 44 than 54?

c. How much more is 65 than 43?

Part 2

a. Nan had some books. She gave away 93 books. She ended up with 78 books. How many books did Nan start with?

b. A store had 215 dolls. Then the store bought some more dolls. The store ended up with 368 dolls. How many dolls did the store buy?

c. A woman weighed 162 pounds. She lost some weight. She ended up weighing 120 pounds. How many pounds did the woman lose?

d. There were 15 dogs in the kennel. 19 more dogs came to stay. How many dogs ended up in the kennel?

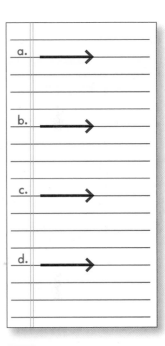

Lesson 60

Part 3

a. Ben was 27 years older than his son. Ben was 39 years old. How many years old was his son?

b. Ann was 11 inches shorter than her mother. Her mother was 60 inches tall. How many inches tall was Ann?

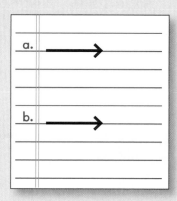

Part 4 Work the problem for **a**. Then copy and work problems **b** and **c**.

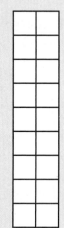

a.

a. ■ x ■ = ■

b. 5 x 6 = ■

c. 4 x 4 = ■

Part 5 Write the letter or letters for each shape: **Cu, RP, Sp, P.**

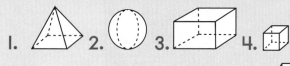

1. 2. 3. 4.

5. 6. 7. 8.

1.	2.	3.	4.
5.	6.	7.	8.

Part 6

a. N is 34 less than R.
R is 62. What number is N?

b. K is 56 more than V.
K is 82. What number is V?

Lesson

a. How much less is 13 than 21?

b. How much less is 52 than 69?

c. How much more is 28 than 18?

Independent Work

Part 2

a. Rita had 248 rocks. She threw away some rocks. She ended up with 135 rocks. How many rocks did she throw away?

b. Fred had 56 rocks. Then he found another 120 rocks. How many rocks did he end up with?

c. You have some pins. You buy 435 pins. You end up with 867 pins. How many pins did you start with?

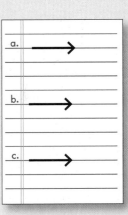

Part 3

a. P is 18 less than T.
P is 28. What number is T?

b. R is 44 less than Y.
Y is 88. What number is R?

c. K is 36 more than F.
K is 82. What number is F?

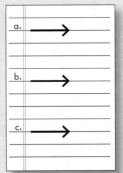

Part 4

a. The green truck was 19 feet longer than the red truck. The red truck was 32 feet long. How many feet long was the green truck?

b. Jim was 5 inches shorter than Helen. Helen was 48 inches tall. How many inches tall was Jim?

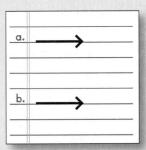

Lesson 62

Part 1

a. How much less is 76 than 96?

b. How much more is 80 than 58?

c. How much more is 95 than 83?

d. How much less is 96 than 99?

Part 2

a. $\dfrac{G \quad M}{} V$

G = 10
V = 12

b. $\dfrac{J \quad L}{} N$

L = 8
J = 30

c. $\dfrac{Z \quad Y}{} X$

Y = 3
X = 7

d. $\dfrac{R \quad T}{} W$

R = 9
T = 6

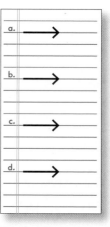

Part 3

a.

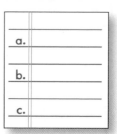

b.

c.

Connecting Math Concepts

Lesson 62

Part 4

a. Joe had some money. He spent 35 dollars and ended up with 129 dollars. How many dollars did Joe have to begin with?

b. Terry had 85 dollars. She spent some money. She has 41 dollars left. How many dollars did Terry spend?

c. Kelly had 21 pieces of wood. She got some more wood. She ended up with 78 pieces of wood. How many pieces of wood did Kelly get?

d. Peggy read 28 pages in her book. Then she read 13 more pages. How many pages did she end up reading?

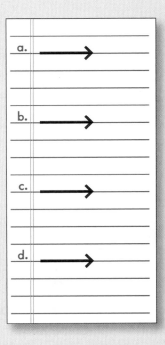

Part 5

Write the problem for **a** and work it. Then copy and work problems **b** and **c**.

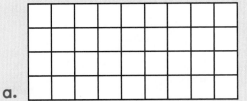

a.

a. $\blacksquare \times \blacksquare = \blacksquare$

b. $5 \times 5 = \blacksquare$

c. $9 \times 5 = \blacksquare$

Part 6

a. Brenda is 26 years older than her son. Her son is 26 years old. How many years old is Brenda?

b. The black dog weighs 22 pounds less than the white dog. The black dog weighs 76 pounds. How many pounds does the white dog weigh?

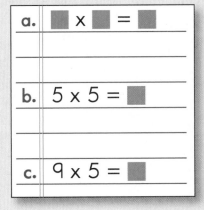

Lesson 63

Part 1

a.
$$\overrightarrow{Q \quad R}P$$
R = 15
P = 25

b.
$$\overrightarrow{T \quad M}V$$
M = 11
T = 25

c.
$$\overrightarrow{F \quad N}Z$$
F = 40
Z = 60

d.
$$\overrightarrow{Y \quad Z}x$$
x = 42
y = 30

a. →
b. →
c. →
d. →

Part 2

a. How much older is car X than car Y?

b. How much shorter is Henry than Fran?

c. How much taller is house V than house K?

a. →
b. →
c. →

Part 3

a.

b.

c.

a.
b.
c.

Lesson

Part 4

a. Rob started with 24 eggs. He broke some eggs. He ended up with 14 eggs. How many eggs did he break?

b. Pat started out with some shells. He found 12 shells and ended up with 30 shells. How many shells did he start with?

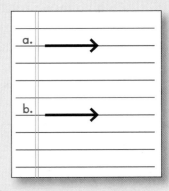

Part 5

a. J is 64 less than T.
   J is 23. What number is T?

b. P is 57 less than M.
   M is 64. What number is P?

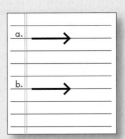

Part 6

Write the problem for **a** and work it. Then copy and work problems **b** and **c**.

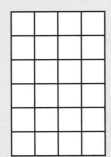

a.

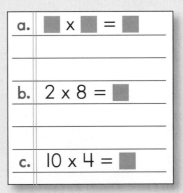

a. ■ x ■ = ■

b. 2 x 8 = ■

c. 10 x 4 = ■

Part 7

a. Will has 22 fewer cards than Matt. Will has 58 cards. How many cards does Matt have?

b. Sue is 26 pounds heavier than Mo. Mo weighs 69 pounds. How many pounds does Sue weigh?

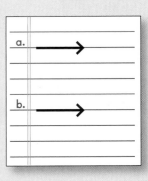

Lesson 64

Part 1

a. How much longer is rope K than rope R?

b. How much younger is Adam than Sue?

c. How much shorter is the house than the barn?

d. How much heavier is the cow than the goat?

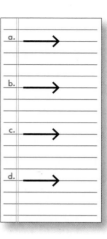

Part 2

a. $\xrightarrow{\quad v \quad m \quad} k$

m = 41

v = 30

b. $\xrightarrow{\quad t \quad s \quad} e$

s = 7

e = 14

c. $\xrightarrow{\quad r \quad n \quad} q$

r = 20

q = 70

d. $\xrightarrow{\quad w \quad y \quad} z$

w = 9

y = 7

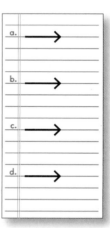

Connecting Math Concepts

Lesson

Part 3 Write the dollars and cents number: $■■.■■

a.

| a. |
| b. |

b.

Part 4

a. Mary had 52 dollars. She spent 16 dollars. How many dollars did she end up with?

b. Bill had 42 stamps. He bought more stamps. He ended up with 95 stamps. How many stamps did he buy?

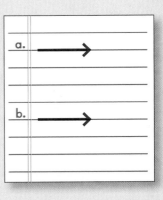

Part 5

a. Y is 36 more than B.
B is 26. What number is Y?

b. J is 26 less than P.
P is 36. What number is J?

Lesson 65

Part 1

a. Tom is 16 years old. Mary is 47 years old. How many years older is Mary than Tom?

b. Sam ran 18 miles. Fran ran 6 miles. How many more miles did Sam run than Fran?

c. Hillary is 67 inches tall. Ann is 50 inches tall. How many inches shorter is Ann than Hillary?

d. Vern had 6 blue shirts. Greg had 11 blue shirts. How many more blue shirts did Greg have than Vern had?

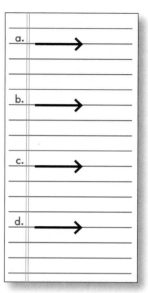

Independent Work

Part 2

Copy each number family. Put 2 numbers in each family. Figure out the missing number.

a. $\xrightarrow[]{\text{T} \quad \text{B}}K$
   B = 8
   K = 18

b. $\xrightarrow[]{\text{R} \quad \text{Y}}F$
   R = 45
   Y = 23

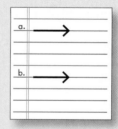

Part 3

Write the problem for **a** and work it. Then copy and work problems **b** and **c**.

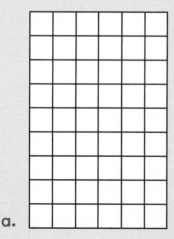

a.

a. ■ x ■ = ■

b. 5 x 3 = ■

c. 2 x 5 = ■

Connecting Math Concepts

Lesson 65

Part 4

a. A man starts out with 29 boxes. He makes some more boxes. He ends up with 40 boxes. How many boxes did he make?

b. Mary starts out with some dollars. She spends 18 dollars. She ends up with 14 dollars. How many dollars did she start with?

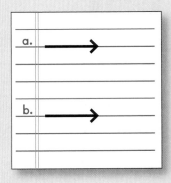

Part 5 Write the dollars and cents number: $■■.■■

a.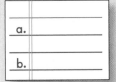

b.

Part 6

a. The car was 19 feet shorter than the boat. The boat was 41 feet long. How many feet long was the car?

b. Jan is 29 years younger than her mother. Jan is 33 years old. How many years old is her mother?

c. The red car is 39 years older than the blue car. The blue car is 23 years old. How many years old is the red car?

Lesson 66

Part 1

a. Benny weighs 110 pounds. Linda weighs 137 pounds. How many pounds lighter is Benny than Linda?

b. Rita has 235 nails. George has 25 nails. How many more nails does Rita have than George has?

c. Donna is 11 years old. Benny is 46 years old. How many years younger is Donna than Benny?

d. The elm tree was 46 feet tall. The maple tree was 157 feet tall. How many feet shorter was the elm tree than the maple tree?

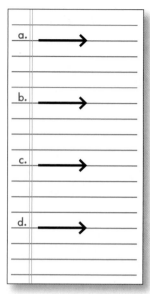

Independent Work

Part 2

Write the problem for **a** and work it. Then copy and work problems **b** and **c**.

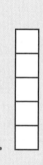

a.

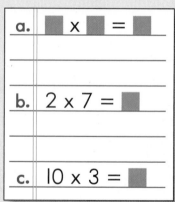

a. ■ x ■ = ■

b. 2 x 7 = ■

c. 10 x 3 = ■

Part 3

Write the letter or letters for each shape: **R, S, P, Cu, RP, Sp.**

a. △

b. (cube)

c. (pyramid)

a.	b.	c.
d.	e.	f.

d. (rectangular prism)

e. (rectangle)

f. (square)

Connecting Math Concepts

Lesson 66

Part 4

a. Emily ran 17 miles farther than her brother ran. Her brother ran 23 miles. How many miles did Emily run?

b. The tree was 56 feet taller than the house. The tree was 75 feet tall. How many feet tall was the house?

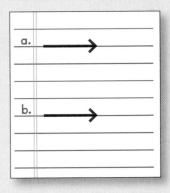

Part 5 Write the dollars and cents number: $■■.■■

a.

b.

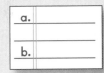

Part 6 Copy each number family. Figure out the missing number.

a.
$$\underset{\text{p} \quad \text{r}}{\longrightarrow}\text{v}$$
r = 131
v = 533

b.
$$\underset{\text{k} \quad \text{m}}{\longrightarrow}\text{n}$$
m = 62
k = 37

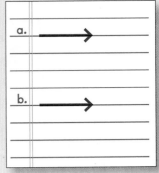

Lesson 67

Part 1

animals   relatives   vehicles   containers   vegetables

a. | truck
   | motorcycle

b. | father
   | sister

c. | potato
   | onion

d. | monkey
   | wolf

e. | box
   | jug

a. ⟶
b. ⟶
c. ⟶
d. ⟶
e. ⟶

Part 2

a. Jane ran 11 miles farther than Ginger ran. Ginger ran 9 miles. How many miles did Jane run?

b. Jane rode a bike for 52 miles. Ginger rode for 36 miles. How many more miles did Jane ride than Ginger?

c. Bill is 11 inches taller than Fran. Bill is 73 inches tall. How many inches tall is Fran?

d. Al weighs 135 pounds. Janice weighs 110 pounds. How many pounds heavier is Al than Janice?

a. ⟶
b. ⟶
c. ⟶
d. ⟶

Connecting Math Concepts

Lesson

Part 3 Write the problem for **a** and work it. Then copy and work problems **b** and **c**.

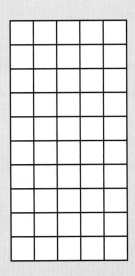

a.

a. ■ x ■ = ■

b. 10 x 5 = ■

c. 2 x 5 = ■

Part 4

a. Mr. Brown starts out with 18 dogs. Then he buys some more dogs. He ends up with 30 dogs. How many dogs did he buy?

b. The train starts out with 189 people on it. Then 21 more people got on the train. How many people ended up on the train?

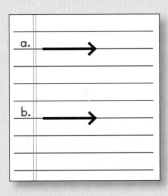

Part 5 Copy each number family. Figure out the missing number.

a. $\xrightarrow[\quad]{p \qquad q} r$

q = 70
r = 92

b. $\xrightarrow[\quad]{x \qquad y} z$

x = 25
z = 67

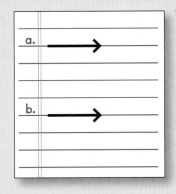

Lesson 68

Part 1

a. $3 + \blacksquare = 9$

d. $2 + \blacksquare = 8$

b. $8 + \blacksquare = 16$

e. $3 + \blacksquare = 11$

c. $10 + \blacksquare = 12$

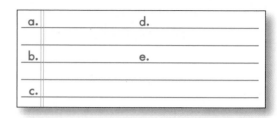

Part 2

a. An elm tree was 15 feet shorter than a pine tree. The pine tree was 84 feet tall. How many feet tall was the elm tree?

b. Alice ran 16 miles farther than Rita. Rita ran 26 miles. How many miles did Alice run?

c. Fran ran 17 miles. Rita ran 6 miles. How many more miles did Fran run than Rita ran?

d. Don had 78 more eggs than George had. George had 220 eggs. How many eggs did Don have?

e. An elm tree was 15 feet tall. A pine tree was 84 feet tall. How many feet shorter was the elm tree than the pine tree?

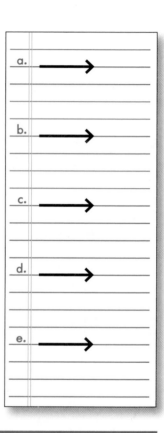

Part 3

relatives   tools   food   containers   bugs

a. meat
pizza

b. hammer
rake

c. ant
spider

d. glass
sack

e. brother
cousin

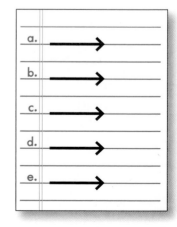

Connecting Math Concepts

Lesson

Independent Work

Part 4

a. Mrs. Briggs drove her car 153 miles in the morning. She drove some more miles in the afternoon. She ended up driving 306 miles in all. How many miles did she drive in the afternoon?

b. Henry started out with 310 nails. Then he bought 500 more nails. How many nails did he end up with?

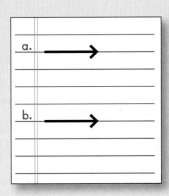

Part 5 Write the letter or letters for each shape: R, T, C, P, Cu, RP, Sp.

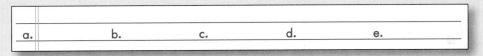

a.   b.   c.   d.   e.

Part 6 Write the dollars and cents number: $■■.■■

a.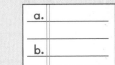

b.

Part 7

a. The green house was 32 years older than the yellow house. The green house was 39 years old. How old was the yellow house?

b. The green train was 89 feet shorter than the blue train. The green train was 503 feet long. How long was the blue train?

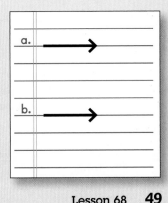

Connecting Math Concepts

Lesson 68 **49**

Lesson 69

Part 1

a. $20 + \blacksquare = 28$ d. $10 + \blacksquare = 15$

b. $9 + \blacksquare = 18$ e. $4 + \blacksquare = 8$

c. $2 + \blacksquare = 10$

a.		d.
b.		e.
c.		

Part 2

a. Rita had 56 feet of string. Al had 16 feet of string. How many feet shorter was Al's string than Rita's string?

b. Fran walked 12 more miles than Al walked. Al walked 14 miles. How many miles did Fran walk?

c. Donna weighed 22 pounds less than Eric weighed. Donna weighed 150 pounds. How many pounds did Eric weigh?

d. A car was 23 feet long. A truck was 59 feet long. How many feet longer was the truck than the car?

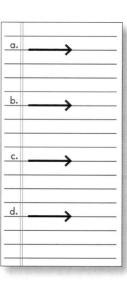

Part 3

a.
boys
children
girls

b.
cats
dogs
pets

c.
food
nachos
hamburger

d.
bike
vehicle
car

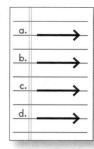

Independent Work

Part 4 Write the dollars and cents number: $\blacksquare.\blacksquare\blacksquare$

a.

a.
b.

b.

Lesson

Part 5

a. Jan started out with some baskets. Then she gave away 22 baskets. She ends up with 70 baskets. How many baskets did she start out with?

b. There were 182 bricks in a pile. A truck brought 711 more bricks. How many bricks ended up in the pile?

c. A farmer starts out with 13 cows. Then he buys some cows. He ends up with 69 cows. How many cows did he buy?

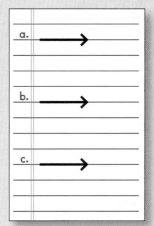

Part 6 Write the problem for **a** and work it. Then copy and work problems **b** and **c**.

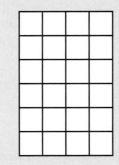

a.

a. ▮ x ▮ = ▮

b. 9 x 6 = ▮

c. 5 x 6 = ▮

Part 7 Copy each number family. Figure out the missing number.

a.
$$\xrightarrow[\ \ \ \ \ ]{p\qquad r}t$$
p = 18
t = 81

b.
$$\xrightarrow[\ \ \ \ \ ]{k\qquad v}p$$
p = 45
v = 34

c.
$$\xrightarrow[\ \ \ \ \ ]{m\qquad j}y$$
j = 188
m = 109

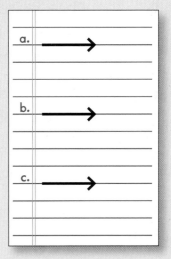

Lesson 70

Part 1

a.
bananas
oranges
fruit

b.
people
women
men

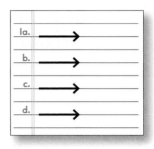

c.
drill
hammer
tool

d.
stores
buildings
houses

Part 2

a. $16 - 8 = \underline{}$

b. $2 + \underline{} = 12$

c. $3 + \underline{} = 10$

d. $17 - 10 = \underline{}$

e. $20 + \underline{} = 28$

f. $12 - 6 = \underline{}$

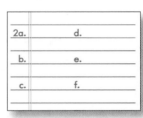

Part 3

a. Billy is 22 years old. Albert is 57 years old. How many years younger is Billy than Albert?

b. The white fence is 15 feet longer than the blue fence. The blue fence is 45 feet long. How many feet long is the white fence?

c. Helen has 20 fewer stamps than Mary. Helen has 68 stamps. How many stamps does Mary have?

d. The stool is 30 inches tall. The table is 18 inches tall. How many inches taller is the stool than the table?

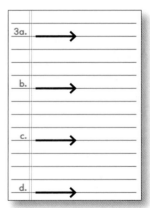

Connecting Math Concepts

Lesson 70

Part 4 Write the problem for **a** and work it. Then copy and work problems **b** and **c**.

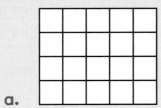

a.

4a. ■ x ■ = ■

b. 9 x 6 = ■

c. 4 x 6 = ■

Part 5 Write the dollars and cents number: $■■.■■

a.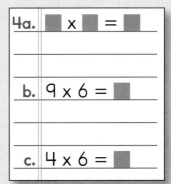

5a.

b.

b.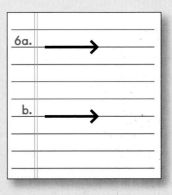

Part 6

a. Dan started out with 16 blue stones. Then he found some more blue stones. He ended up with 88 blue stones. How many blue stones did he find?

6a. ⟶

b. At first, there were 67 birds on the lake. Then 32 more birds landed on the lake. How many birds ended up on the lake?

b. ⟶

Part 7 Write the letter or letters for each shape: R, S, P, Cu, RP, Sp.

7a.   b.   c.   d.

a.   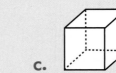 b.   c.   d.

Lesson

a. children girls boys	b. motorcycles airplanes vehicles	c. building garage school	d. breakfast meal lunch

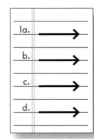

Independent Work

Part 2 Write the complete equation for each problem.

a. $8 + \blacksquare = 28$ d. $9 - 3 = \blacksquare$

b. $30 + \blacksquare = 40$ e. $6 + \blacksquare = 12$

c. $19 - 10 = \blacksquare$ f. $11 - 9 = \blacksquare$

2a.	d.
b.	e.
c.	f.

Part 3

a. Rhonda started out with 178 papers. She sold some papers. She ended up with 120 papers. How many papers did she sell?

b. The train started out with some people on it. 103 people got off the train. The train ended up with 59 people on it. How many people started out on the train?

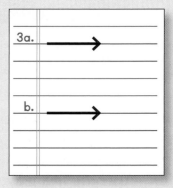

Part 4 Write the problem for **a** and work it. Then copy and work problems **b** and **c**.

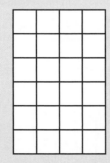

a.

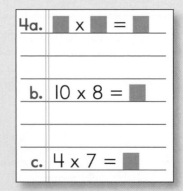

4a. $\blacksquare \times \blacksquare = \blacksquare$

b. $10 \times 8 = \blacksquare$

c. $4 \times 7 = \blacksquare$

Lesson

Copy each number family. Figure out the missing number.

a. $\xrightarrow[\substack{p = 36 \\ r = 44}]{p \quad r \,}j$

b. $\xrightarrow[\substack{t = 150 \\ n = 38}]{m \quad n \,}t$

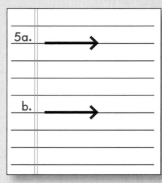

Part 6

a. Wendy is 28 years old. Bryan is 42 years old. How many years older is Bryan than Wendy?

b. The cat weighs 8 pounds. The cat is 18 pounds lighter than the dog. How many pounds does the dog weigh?

c. The brick house is 15 feet taller than the wood house. The wood house is 25 feet tall. How many feet tall is the brick house?

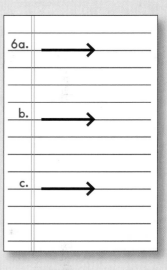

Lesson 72

a. 14 girls
16 children
How many boys?

b. 9 trucks
How many vehicles?
7 cars

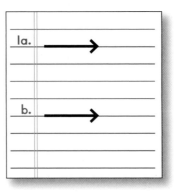

Part 2

a. $\blacksquare + 105 = 215$

b. $512 + \blacksquare = 830$

c. $\blacksquare + 92 = 99$

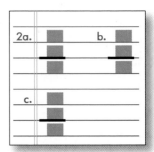

Independent Work

Part 3 Write the problem for **a** and work it. Then copy and work problems **b** and **c.**

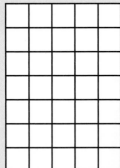

a.

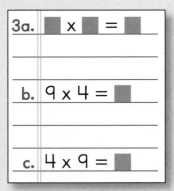

3a. $\blacksquare \times \blacksquare = \blacksquare$

b. $9 \times 4 = \blacksquare$

c. $4 \times 9 = \blacksquare$

Part 4 Write the complete equation for each problem.

a. $6 + \blacksquare = 12$

d. $7 + 7 = \blacksquare$

b. $10 - 2 = \blacksquare$

e. $2 + 8 = \blacksquare$

c. $10 + \blacksquare = 18$

f. $20 + \blacksquare = 25$

<table>
<tr><td>4a.</td><td>d.</td></tr>
<tr><td>b.</td><td>e.</td></tr>
<tr><td>c.</td><td>f.</td></tr>
</table>

Connecting Math Concepts

Lesson 72

Part 5 Write the dollars and cents number: $■■.■■

a.

5a.	
b.	

b.

Part 6 Copy each number family. Figure out the missing number.

a.
$$\xrightarrow[\quad m \quad\quad r \quad]{} v$$
v = 425
r = 20

b.
$$\xrightarrow[\quad b \quad\quad j \quad]{} t$$
j = 136
b = 124

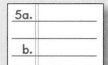

6a.	→
b.	→

Part 7

a. Jerry is 34 years old. Fran is 45 years old. How many years older is Fran than Jerry?

b. The red bag weighs 78 pounds. The white bag weighs 39 pounds. How many pounds more does the red bag weigh?

c. Al's car is 11 years older than Mary's car. Mary's car is 20 years old. How old is Al's car?

7a.	→
b.	→
c.	→

Lesson 73

Part 1

a. How many lunches?
10 meals
3 dinners

b. 18 women
9 men
How many people?

c. 16 vehicles
How many planes?
6 boats

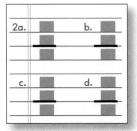

Part 2

a. $132 + \blacksquare = 146$ b. $\blacksquare + 58 = 76$

c. $\blacksquare + 110 = 230$ d. $57 + \blacksquare = 89$

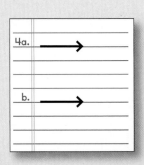

Independent Work

Part 3

Write the problem for **a** and work it. Then copy and work problems **b** and **c**.

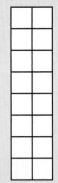

a.

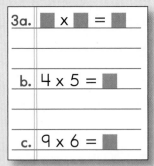

Part 4

a. Bill had 95 dollars. He spent 45 dollars. How many dollars did he end up with?

b. Dan had some marbles. After he bought 26 more, he ended up with 198 marbles. How many marbles did he start with?

Connecting Math Concepts

Lesson 73

Part 5 Write the dollars and cents number: $■■.■■

a.

5a. _____
b. _____

b.

Part 6 Write the complete equation for each problem.

a. $30 + \blacksquare = 39$ d. $6 + \blacksquare = 9$

b. $16 - 8 = \blacksquare$ e. $10 - 5 = \blacksquare$

c. $11 - 9 = \blacksquare$ f. $10 + \blacksquare = 15$

6a.	d.
b.	e.
c.	f.

Part 7

a. The pig weighs 42 pounds more than the dog. The dog weighs 29 pounds. How many pounds does the pig weigh?

b. Tammy weighs 17 pounds less than her brother. Tammy weighs 72 pounds. How many pounds does her brother weigh?

c. The blue house is 38 years old. The white house is 12 years old. How many years older is the blue house?

7a. ⟶

b. ⟶

c. ⟶

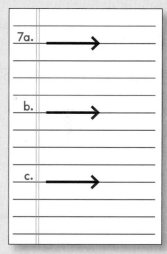

Lesson 74

a. 12 brooms
    How many tools?
    10 screwdrivers

b. How many women?
    50 people
    20 men

c. 15 birds
    3 robins
    How many ducks?

d. 11 rabbits
    23 pets
    How many cats?

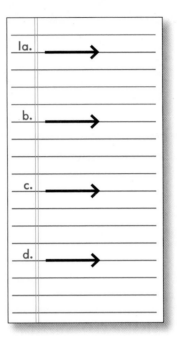

Part 2

a. $125 - \blacksquare = 105$ b. $96 - \blacksquare = 64$

c. $180 - \blacksquare = 22$

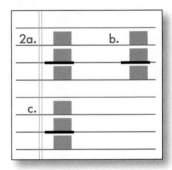

Independent Work

Part 3

Write the problem for **a** and work it. Then copy and work problems **b** and **c.**

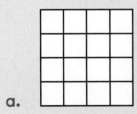

a.

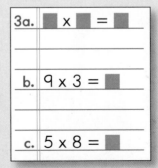

3a. $\blacksquare \times \blacksquare = \blacksquare$

b. $9 \times 3 = \blacksquare$

c. $5 \times 8 = \blacksquare$

Lesson 74

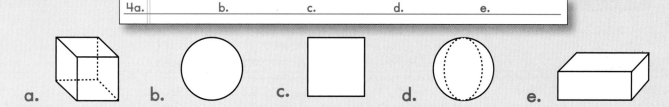

Part 4
Write the letter or letters for each shape: **R, S, T, C, P, Cu, RP, Sp.**

4a. _____ b. _____ c. _____ d. _____ e. _____

a. b. c. d. e.

Part 5
Copy each number family. Figure out the missing number.

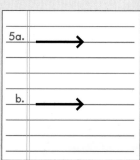

a. $\xrightarrow[\quad]{m \quad f} z$

z = 76
f = 28

b. $\xrightarrow[\quad]{k \quad j} v$

j = 137
k = 43

Part 6

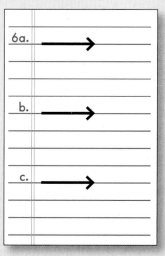

a. Heidi is 11 inches shorter than her dad. Her dad is 80 inches tall. How tall is Heidi?

b. Dawn ran 13 miles last week. Audrie ran 20 miles last week. How many more miles did Audrie run?

c. The old stove weighs 42 pounds less than the new stove. The new stove weighs 151 pounds. How much does the old stove weigh?

Lesson 75

Part 1

a. 8 pears
How many fruits?
3 cherries

b. 19 children
How many boys?
9 girls

c. 10 robins
17 birds
How many crows?

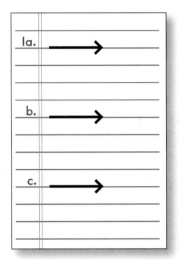

Part 2

a. $\blacksquare + 32 = 85$

b. $98 - \blacksquare = 38$

c. $15 + \blacksquare = 87$

d. $999 - \blacksquare = 663$

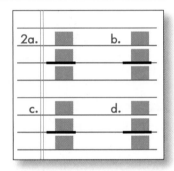

Independent Work

Part 3

a. Tom is 19 years younger than his brother. His brother is 51. How old is Tom?

b. Mr. Brown is 73 inches tall. His wife is 64 inches tall. How much taller is Mr. Brown than his wife?

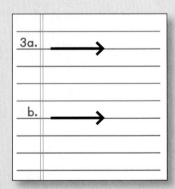

Lesson 75

Part 4 Write the problem for **a** and work it. Then copy and work problems **b–d**.

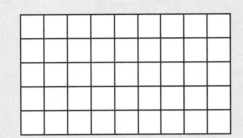

a.

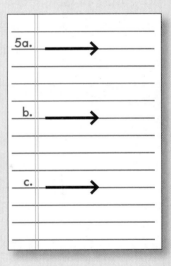

4a. ☐ x ☐ = ☐

b. 4 x 5 = ☐

c. 9 x 6 = ☐

d. 10 x 4 = ☐

Part 5

a. Randy started out with 218 dollars. Then he earned 78 dollars. How much money did he end up with?

b. Barb started out with some dolls. She gave away 28 of those dolls. She ended up with 28 dolls. How many dolls did she start with?

c. A train started out with 156 passengers. Some passengers got off the train. The train ended up with 128 passengers. How many passengers got off the train?

5a. ⟶

b. ⟶

c. ⟶

Lesson 76

Part 1

a. How many turkeys?
30 birds
28 crows

b. 15 vehicles
5 cars
How many trucks?

c. 20 hospitals
How many buildings?
8 stores

d. How many children?
11 boys
34 girls

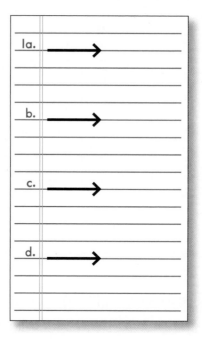

Part 2

a. $126 - \blacksquare = 113$

d. $425 - \blacksquare = 122$

b. $15 + \blacksquare = 97$

e. $92 - \blacksquare = 71$

c. $\blacksquare + 50 = 80$

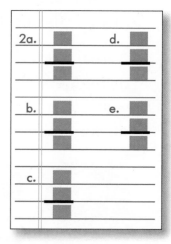

Part 3

a. 57

b. 84

c. 22

d. 46

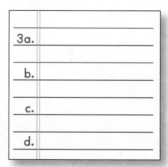

Lesson 76

Part 4

a. Barb started out with 27 buttons. Then she found 33 buttons. How many buttons did she end up with?

b. A train started out with 262 passengers. Then some passengers got off the train. The train ended up with 156 passengers. How many passengers got off the train?

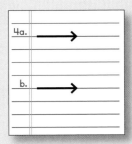

Part 5 Write the letter or letters for each shape: R, S, T, C, P, Cu, RP, Sp.

a. b. c. d.

e. f. g.

5a.	b.	c.	d.
e.	f.	g.	

Part 6 Copy each number family. Figure out the missing number.

a.

$$y = 26$$
$$z = 52$$

b.

$$p = 52$$
$$q = 36$$

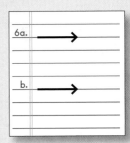

Part 7

a. Milly had 133 more dollars than Dan. Dan had 129 dollars. How many dollars did Milly have?

b. The black boat was 144 feet long. The red boat was 37 feet long. How much longer was the black boat than the red boat?

c. Jerry weighed 228 pounds. His sister weighed 116 pounds. How many more pounds did Jerry weigh than his sister?

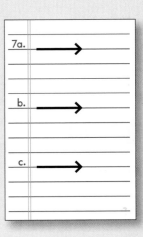

Lesson 77

Part 1

a.	$8 - 4 = \blacksquare$		f.	$9 - 5 = \blacksquare$
b.	$5 + 4 = \blacksquare$		g.	$9 - 4 = \blacksquare$
c.	$9 - 4 = \blacksquare$		h.	$10 - 4 = \blacksquare$
d.	$10 - 6 = \blacksquare$		i.	$4 + 6 = \blacksquare$
e.	$4 + 4 = \blacksquare$		j.	$4 + 5 = \blacksquare$

1a.	f.
b.	g.
c.	h.
d.	i.
e.	j.

Part 2

a. How many pounds did Al buy?

b. How many feet longer is the red truck?

c. How many teeth did the dentist fix?

d. How many hours did the dog sleep?

Part 3

a. How many saws?
80 tools
55 hammers

b. 32 dolls
How many toys?
27 puzzles

c. 40 hotels
89 buildings
How many stores?

d. 75 animals
How many horses?
53 cows

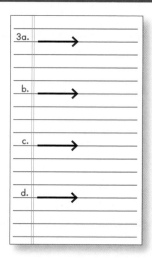

Part 4

a.	68		d.	86
b.	44		e.	12
c.	17			

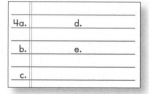

Connecting Math Concepts

Lesson 77

Part 5 Work the column problem for each item.

a. $123 - \blacksquare = 102$ d. $58 - \blacksquare = 45$

b. $48 + \blacksquare = 76$ e. $\blacksquare + 180 = 291$

c. $\blacksquare + 20 = 63$

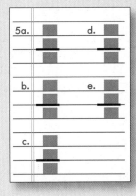

Part 6

a. Hillary had 56 more cookies than Jan had. Jan had 16 cookies. How many cookies did Hillary have?

b. The blue plane was 176 feet long. The blue plane was 16 feet shorter than the red plane. How long was the red plane?

c. Jed's school was 60 feet tall. Jed's school was 45 feet taller than Bernie's school. How tall was Bernie's school?

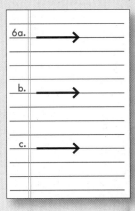

Lesson 77

Part 7

Write the dollars and cents number: $■■.■■

a.

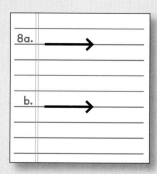

7a. _____

Part 8

Copy each number family. Figure out the missing number.

a. $\xrightarrow{\underline{m \qquad n}} p$

p = 260
n = 154

b. $\xrightarrow{\underline{t \qquad u}} v$

u = 143
v = 190

8a. ————→

b. ————→

Part 9

Write the problem for **a** and work it. Then copy and work problems **b–e**.

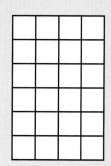

a.

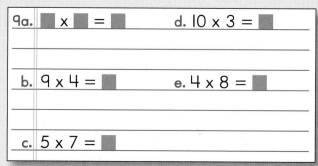

9a. ■ x ■ = ■   d. 10 x 3 = ■

b. 9 x 4 = ■   e. 4 x 8 = ■

c. 5 x 7 = ■

Part 10

a. The cat started out with 66 fleas. Then 38 fleas jumped off the cat. How many fleas ended up on the cat?

b. A train started out with 74 passengers. Then some passengers got on the train. The train ended up with 184 passengers. How many passengers got on the train?

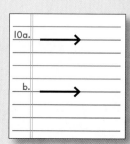

10a. ————→

b. ————→

Lesson 78

Part 1

a. How many feet longer is the boat?

b. How many pounds of food did the goats eat?

c. How many bags did the truck hold?

d. How many people were on the train?

e. How many marbles were on the floor?

Part 2

a. $10 - 4 = \blacksquare$

b. $5 + 4 = \blacksquare$

c. $4 + 6 = \blacksquare$

d. $9 - 4 = \blacksquare$

e. $9 - 5 = \blacksquare$

f. $6 + 4 = \blacksquare$

g. $10 - 6 = \blacksquare$

h. $4 + 5 = \blacksquare$

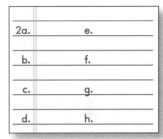

Part 3

a. 39 lunches
How many meals?
21 dinners

b. 20 goldfish
70 rabbits
How many pets?

c. How many carrots?
16 potatoes
22 vegetables

d. 90 people
50 women
How many men?

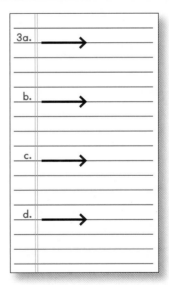

Part 4

a. 29

b. 14

c. 83

d. 76

e. 22

f. 18

Lesson 78

Part 5

a. three hundred seven

b. forty-four

c. fifteen

d. twenty-eight

e. one hundred fifty-six

f. eighty

g. nine hundred ten

5a.	e.
b.	f.
c.	g.
d.	

Independent Work

Part 6

Write the dollars and cents number: $■■.■■

a.

6a.	

Part 7

a. The white house was 52 years older than the red house. The red house was 26 years old. How old was the white house?

b. The big farm had 64 cows. The small farm had 37 cows. How many more cows did the big farm have than the small farm?

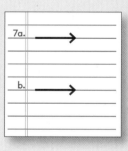

Part 8

Work the column problem for each item.

a. $179 - ■ = 126$ b. $45 + ■ = 80$

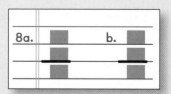

Lesson 78

Part 9 Copy each number family. Figure out the missing number.

a. $\xrightarrow[\text{j = 165}]{\text{j \quad k}} 282$

b. $\xrightarrow[\substack{\text{v = 381} \\ \text{p = 288}}]{\text{p \quad v}} t$

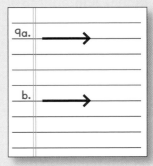

Part 10 Copy and work each problem.

a. $5 \times 10 = \blacksquare$

b. $10 \times 10 = \blacksquare$

c. $4 \times 10 = \blacksquare$

d. $2 \times 10 = \blacksquare$

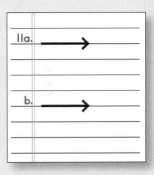

Part 11

a. A plane started out with some passengers on it. Then 206 passengers got off the plane. The plane ended up with 85 passengers. How many passengers started out on the plane?

b. A truck started out with 830 bags on it. Then 264 more bags were put on the truck. How many bags did the truck end up with?

Lesson 79

Part 1

1 cm

5 cm

a. [vertical rectangle]

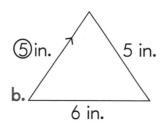
⑤ in.   5 in.
b.   6 in.

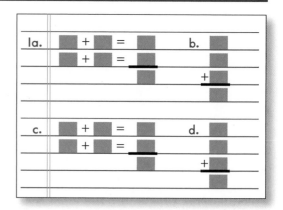
1a. ▢ + ▢ = ▢   b. ▢
▢ + ▢ = ▢   ▢
▢   ▢ +
c. ▢ + ▢ = ▢   d. ▢
▢ + ▢ = ▢   ▢
▢   ▢ +

4 in.

4 in.   [square]

c.

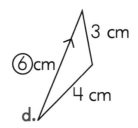
3 cm
⑥ cm
4 cm
d.

Part 2

a. 37   b. 42   c. 74

d. 58   e. 13   f. 86

2a.		b.		c.
d.		e.		f.

Independent Work

Part 3

a. 11 children
   9 boys
   How many girls?

b. How many cats?
   21 pets
   10 dogs

c. 16 lunches
   How many meals?
   20 dinners

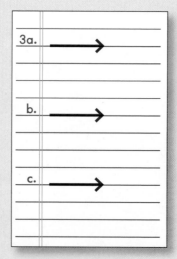
3a. →
b. →
c. →

Lesson 79

Part 4 Write these letters for the shapes: **R, S, T, C, P, Cu, RP, Sp.**

a. ◯   b. ▢   c. ▯   d. △

e. △(pyramid)   f. (rectangular prism)   g. (cube)   h. ◯(sphere)

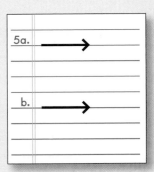

4a.	b.	c.	d.
e.	f.	g.	h.

Part 5

a. The small farm had 86 horses. The small farm had 18 more horses than the big farm. How many horses did the big farm have?

b. The red rock weighed 105 more pounds than the black rock. The black rock weighed 285 pounds. How many pounds did the red rock weigh?

5a.	→
b.	→

Lesson 79

Part 6
Work the column problem for each item.

a. $390 - \blacksquare = 165$ b. $134 + \blacksquare = 367$

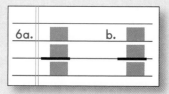

Part 7
Copy each number family. Figure out the missing number.

a. $\dfrac{230 \quad P}{q = 445} \longrightarrow q$

b. $\dfrac{x \qquad y}{\begin{array}{l} z = 807 \\ y = 757 \end{array}} \longrightarrow z$

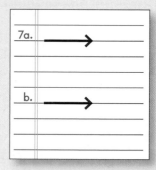

Part 8
Copy and work each problem.

a. $4 \times 7 = \blacksquare$ b. $2 \times 8 = \blacksquare$

c. $2 \times 6 = \blacksquare$ d. $9 \times 10 = \blacksquare$

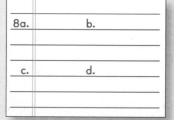

Part 9

a. The duck started out with 472 feathers. The duck ended up with 366 feathers. How many feathers did the duck lose?

b. The truck started out with 564 bags. Then 47 bags fell off the truck. How many bags did the truck end up with?

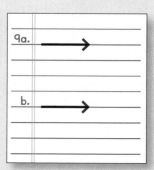

Lesson 80

Part 1 Write the closer tens number.

a. 76   b. 29   c. 43

1a.	b.	c.

Part 2 Find the perimeter for each figure.

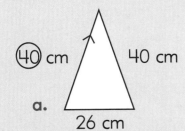

⓵⓪ cm   40 cm

a.

26 cm

7 in.

b.

10 in.

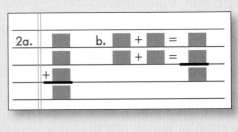

Part 3

a. The boat store started with 76 boats. The store sold 39 boats. How many boats did the store still have?

b. Rob started out with 156 dollars. He spent some dollars. He ended up with 131 dollars. How many dollars did Rob spend?

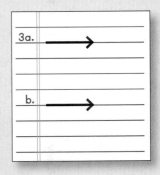

Part 4 Write the dollars and cents number: $■■.■■

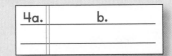

4a.	b.

a.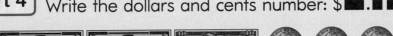

b.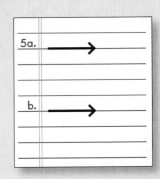

Part 5

a. Barb had 69 cards. Dan had 17 more cards than Barb. How many cards did Dan have?

b. The train had 56 more passengers than the boat. The train had 192 passengers. How many passengers did the boat have?

Lesson 80

Part 6
Copy and work each problem.

a. $2 \times 7 = \blacksquare$ b. $5 \times 7 = \blacksquare$

c. $4 \times 7 = \blacksquare$ d. $9 \times 7 = \blacksquare$

6a.	b.
c.	d.

Part 7

a. How many boats?
14 canoes
12 rafts

b. 18 sheep
How many cows?
30 animals

c. 26 fruits
10 pears
How many apples?

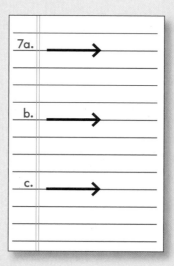

Part 1

a.  7 4

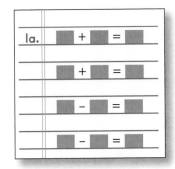

1a. | ▢ + ▢ = ▢ |
| ▢ + ▢ = ▢ |
| ▢ - ▢ = ▢ |
| ▢ - ▢ = ▢ |

Part 2

a.

b.

c.

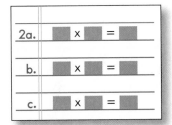

2a. | ▢ × ▢ = ▢ |
b. | ▢ × ▢ = ▢ |
c. | ▢ × ▢ = ▢ |

Part 3

a. 3 + 31 + 9   b. 104 + 5 + 47

c. 28 + 152 + 14

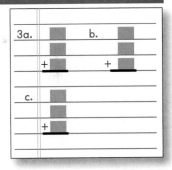

Independent Work

Part 4 Make a number family. Find the missing number.

a. The boat store has 126 motorboats and 19
rowboats. How many more motorboats than
rowboats does the store have?

b. The red tree was 61 feet shorter than the green
tree. The green tree was 190 feet tall. How tall
was the red tree?

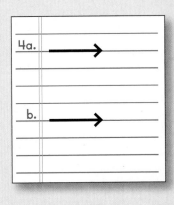

Lesson 81

Part 5 Write the closer tens number.

a. 24   b. 66   c. 52

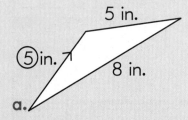

Part 6 Copy and work each problem.

a. $10 \times 7 = \blacksquare$ d. $5 \times 6 = \blacksquare$

b. $4 \times 6 = \blacksquare$ e. $4 \times 8 = \blacksquare$

c. $9 \times 8 = \blacksquare$

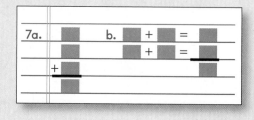

Part 7 Find the perimeter of each figure.

a.
5 in.
⑤ in.
8 in.

b.
10 cm
9 cm

Part 8 Make a number family. Find the missing number.

a.   48 men
How many people?
51 women

b.   How many buildings?
15 schools
25 houses

c.   62 vehicles
14 trucks
How many buses?

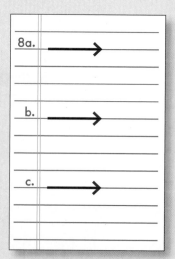

Lesson 82

Part 1

a. $4 + 5 = \blacksquare$

b. $11 - 4 = \blacksquare$

c. $6 + 4 = \blacksquare$

d. $7 + 4 = \blacksquare$

e. $10 - 6 = \blacksquare$

f. $9 - 5 = \blacksquare$

g. $4 + 7 = \blacksquare$

h. $11 - 7 = \blacksquare$

i. $10 - 4 = \blacksquare$

j. $9 - 4 = \blacksquare$

1a.	f.
b.	g.
c.	h.
d.	i.
e.	j.

Part 2

a. fifty-seven

b. three hundred

c. thirteen

d. ninety-eight

e. seventy

f. one hundred eighty-two

2a.	d.
b.	e.
c.	f.

Part 3

a.

b.

c.

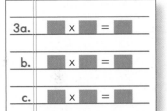

Part 4

a. $5 + 165 + 22$

b. $23 + 443 + 6$

c. $110 + 59 + 28$

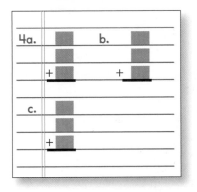

Lesson 82

Part 5 | Make a number family. Find the missing number.

a. The nut store had 240 pounds of nuts in October. The nut store had 351 pounds of nuts in November. How many fewer pounds did the store have in October than November?

b. Liz weighed 39 pounds. Liz weighed 58 fewer pounds than her brother. How many pounds did her brother weigh?

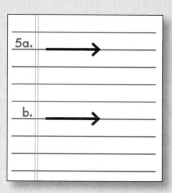

Part 6 | Write the dollars and cents number: $■■.■■

a.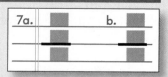

b.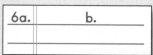

Part 7 | Write and work the column problem for each item.

a. $263 - ■ = 44$ b. $518 + ■ = 637$

Part 8 | Copy and work each problem.

a. $5 \times 9 = ■$

b. $9 \times 3 = ■$

c. $2 \times 9 = ■$

d. $4 \times 4 = ■$

e. $10 \times 6 = ■$

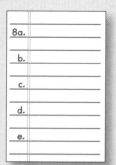

Part 9 | Make a number family. Find the missing number.

a. How many boys?
   18 girls
   40 children

b. 155 carrots
   144 potatoes
   How many vegetables?

c. 80 breakfasts
   How many lunches?
   98 meals

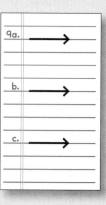

Lesson 83

Part 1

a. Jill is 14 years older than Tim. Jill is 48 years old. How many years old is Tim?

b. There are 48 more brown eggs than white eggs. There are 12 white eggs. How many brown eggs are there?

c. The car was 18 feet long. The car was 22 feet shorter than the boat. How many feet long was the boat?

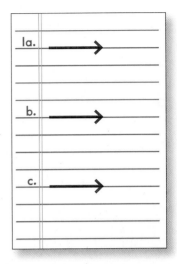

Part 2

a.

b.

c.

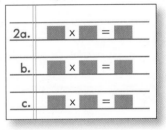

Part 3

a. $4 + 7 = \blacksquare$ **f.** $4 + 5 = \blacksquare$

b. $10 - 4 = \blacksquare$ **g.** $11 - 7 = \blacksquare$

c. $10 - 6 = \blacksquare$ **h.** $9 - 4 = \blacksquare$

d. $9 - 5 = \blacksquare$ **i.** $6 + 4 = \blacksquare$

e. $7 + 4 = \blacksquare$ **j.** $11 - 4 = \blacksquare$

Lesson 83

Independent Work

Part 4   Write each problem in a column and work it.

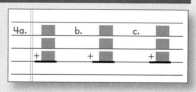

a.   9 + 450 + 107   b.   58 + 8 + 130   c.   102 + 9 + 50

Part 5   Write the closer tens number.

a.   53   b.   97

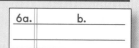

Part 6   Write the dollars and cents number: $■■.■■

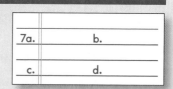

a.

b.   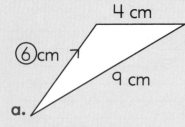 ...

Part 7   Copy and work each problem.

a.   4 × 6 = ■   b.   5 × 6 = ■

c.   2 × 6 = ■   d.   9 × 6 = ■

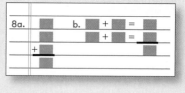

Part 8   Find the perimeter of each figure.

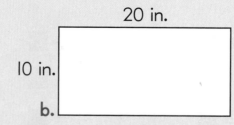

4 cm

(6) cm

9 cm

a.

20 in.

10 in.

b.

Part 9   Write and work the column problem for each item.

a.   9 + ■ = 14   b.   125 − ■ = 16

c.   48 + ■ = 79   d.   15 − 10 = ■

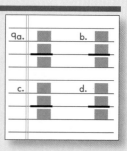

Lesson 84

Part 1

a. There are 18 children in the park. 10 of them are girls. How many of them are boys?

b. There were 72 animals in the field. 23 were squirrels. The rest were mice. How many mice were in the field?

c. 45 men and 25 women were in the gym. How many people were in the gym?

Part 2

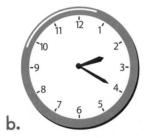  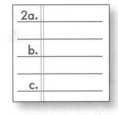

a.   b.   c.

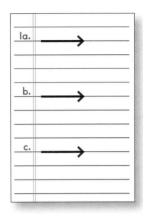

Part 3

a.

b.

c.

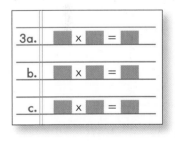

Part 4

a. The red board weighed 45 pounds more than the blue board. The red board weighed 60 pounds. How many pounds did the blue board weigh?

b. Rob is 22 years old. Rob is 6 years older than his sister Sue. How many years old is Sue?

c. Tina is 66 inches tall. Jan is 45 inches tall. How many inches taller is Tina?

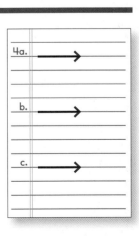

Lesson 84

Independent Work

Part 5
Write the letter or letters for each shape: **R, S, T, C, P, Cu, RP, Sp.**

5a.	b.	c.	d.	e.	f.

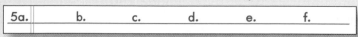

a.   b. 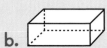 c.   d.   e.   f.

Part 6
Write the answer to each problem.

6a.	b.

a. $60 - 30 = \blacksquare$ b. $160 - 80 = \blacksquare$

Part 7
Write the dollars and cents number: $\$\blacksquare\blacksquare.\blacksquare\blacksquare$

7a.	b.

a.

b.

Part 8
Copy and work each problem.

8a.	b.
c.	d.

a. $2 \times 5 = \blacksquare$ b. $4 \times 7 = \blacksquare$

c. $5 \times 8 = \blacksquare$ d. $9 \times 5 = \blacksquare$

Part 9
Make a number family. Find the missing number.

a. 61 people
32 women
How many men?

b. How many cars?
180 vehicles
73 trains

c. 28 balls
How many toys?
48 jump ropes

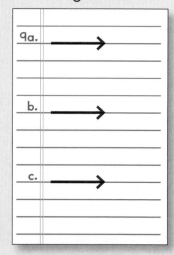

Connecting Math Concepts

Lesson 85

Part 1

a. There were trucks and cars in a parking lot. 14 of the vehicles were trucks. 66 of the vehicles were cars. How many vehicles were in the parking lot?

b. There were apples and pears in a fruit bowl. There were 18 pieces of fruit in all. There were 12 pears. How many apples were there?

c. Rob counted houses and stores on his way home. He counted 73 houses. He counted 80 buildings. How many stores did he count?

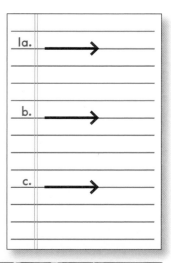

1a. →

b. →

c. →

Part 2

2a. | b. | c.

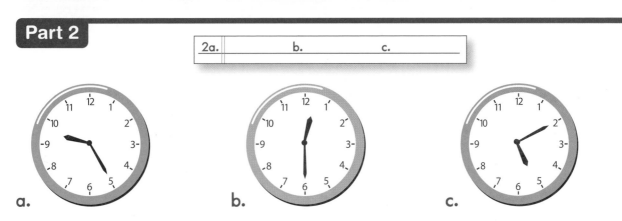

a.   b.   c.

Lesson 85

Part 3

a. The train started out with 82 people on it. Then some people got off. There are now 60 people on the train. How many people got off?

b. Mike had lots of trains. He sold 28 trains. He ended up with 42 trains. How many trains did he start with?

c. Carla had 14 trees in her yard. She planted another 59 trees. How many trees does she have now?

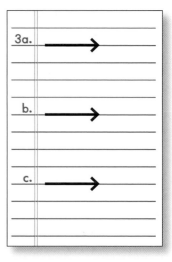

Independent Work

Part 4

Work the times problem for each row.

a.

b.

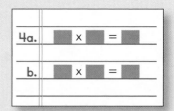

Part 5

Write and work the column problem for each item.

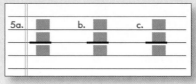

a. $17 + \blacksquare = 24$ b. $\blacksquare + 652 = 678$ c. $260 - \blacksquare = 43$

Part 6

Write 4 facts for each family.

a. $\underset{\longrightarrow}{6 \quad 4} \blacksquare$

b. $\underset{\longrightarrow}{7 \quad 4} \blacksquare$

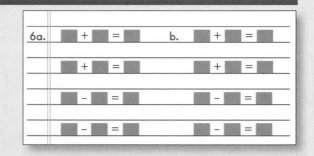

Lesson 86

Part 1

a. There are 36 children in the park. 15 of them are boys. How many are girls?

b. Bill has 19 hammers and 37 wrenches in a bag. How many tools does Bill have in his bag?

c. Jerry had 19 brown cows and some white cows. He had 33 cows in all. How many white cows did he have?

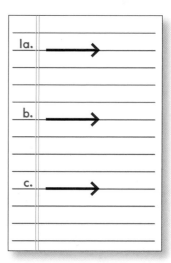

Part 2

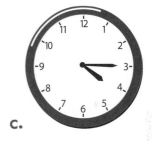

a.   **b.**   **c.**

Part 3

a. $10 - 5 =$ ■   **f.** $11 - 6 =$ ■

b. $6 + 5 =$ ■   **g.** $11 - 5 =$ ■

c. $11 - 5 =$ ■   **h.** $12 - 5 =$ ■

d. $12 - 7 =$ ■   **i.** $5 + 7 =$ ■

e. $5 + 5 =$ ■   **j.** $5 + 6 =$ ■

Lesson 86

Part 4

a. Rob started out with 38 toys. He gave away 14 toys. How many toys did he still have?

b. Jane had some coins. She spent 49 of those coins. She ended up with 25 coins. How many coins did she start with?

c. Cara had 72 eggs in her basket. She dropped and broke some eggs. She has 26 eggs left. How many eggs did she break?

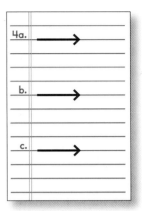

Independent Work

Part 5

a.

b.

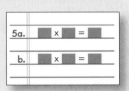

Part 6

Write each problem in a column and work it.

a. 273 + 576   **b.** 15 + 35 + 31

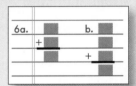

Part 7

Find the perimeter of each figure.

9 in.

9 in.

a.

10 cm

⑥cm

12 cm

b.

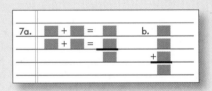

Part 8

Write the answer to each problem.

a. 20 + 60 = ■   **b.** 60 + 30 = ■

c. 10 + 120 = ■   **d.** 70 + 20 = ■

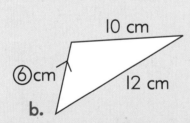

Lesson 87

Part 1

a. A farmer had 88 goats. 29 of the goats were black. The rest were white. How many white goats did he have?

b. There are 74 cars on a lot. 45 of those cars are new. The rest are used. How many used cars are there?

c. There were 83 women in the mall and 86 men in the mall. How many people were in the mall?

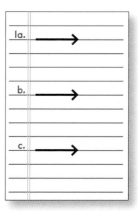

Part 2

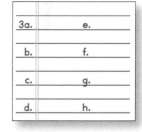

2a.　　　b.　　　c.

a.　　　b.　　　c.

Part 3

a. 12 − 5 = ■

b. 6 + 5 = ■

c. 5 + 7 = ■

d. 11 − 5 = ■

e. 11 − 6 = ■

f. 7 + 5 = ■

g. 12 − 7 = ■

h. 5 + 6 = ■

3a.　　　e.
b.　　　f.
c.　　　g.
d.　　　h.

Part 4

a. Bob started out with 240 cards in his collection. After he bought more cards, he had 286 cards. How many cards did Bob buy?

b. Pat had 125 dollars. He saved another 65 dollars. How many dollars did he end up with?

c. At the start of March, there were 87 cows on the farm. After some calves were born, there were 94 cows on the farm. How many calves were born?

Lesson 87

Independent Work

Part 5

a.

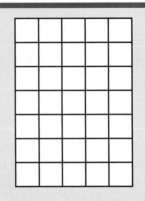

b.

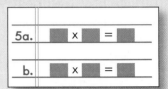

5a. ▦ x ▦ = ▦

b. ▦ x ▦ = ▦

Part 6

Write the dollars and cents number: $■■.■■

6a.   b.

a.

b.

Part 7

Write the column problem for each item.

7a.   b.   c.

a. ■ + 249 = 759

b. 329 + ■ = 685

c. 440 − ■ = 335

Part 8

Write 4 facts for the family.

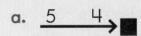

a. 5   4 → ■

b. 7   4 → ■

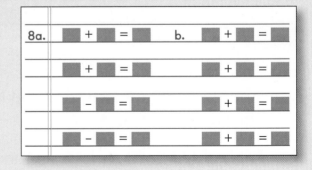

8a. ▦ + ▦ = ▦   b. ▦ + ▦ = ▦

▦ + ▦ = ▦   ▦ + ▦ = ▦

▦ − ▦ = ▦   ▦ + ▦ = ▦

▦ − ▦ = ▦   ▦ + ▦ = ▦

Lesson 88

Part 1

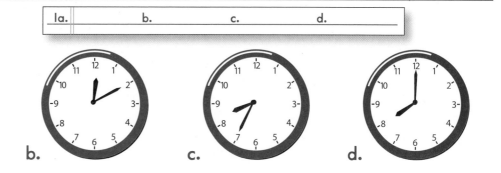

| 1a. | b. | c. | d. |

a.   b.   c.   d.

Independent Work

Part 2

Work each problem. Show the whole answer with a number and unit name.

a. There were 97 birds on the lake. 18 of them were swans. The rest were loons. How many loons were on the lake?

b. 52 men and 38 women signed up for the fun run. How many adults signed up for the fun run?

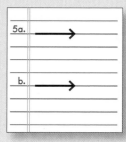

Part 3

Write each problem in a column and work it.

a. 648 + 40 + 303   b. 26 + 46 + 20

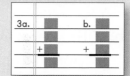

Part 4

Find the perimeter of each figure.

5 cm

3 cm

a.

10 in.

⑨in.

b. 15 in.

Part 5

Work each problem. Show the whole answer with a number and unit name.

a. Hazel had 38 more marbles than Jake. Hazel had 47 marbles. How many marbles did Jake have?

b. The brown horse was 38 months younger than the white horse. The brown horse was 68 months old. How many months old was the white horse?

Lesson 89

Part 1

1a.   b.   c.   d.

a.

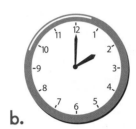
b.

c.

d.

Independent Work

Part 2 Work each problem. Show the whole answer with a number and unit name.

a. The train started out with some passengers. Then 78 passengers got on the train. The train ended up with 290 passengers. How many passengers did the train start out with?

b. The children started out with 123 bottles. Then they found 49 bottles. How many bottles did they end up with?

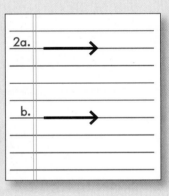

Part 3 Find the perimeter of each figure.

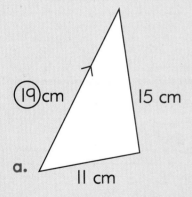
a.

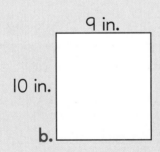
b.

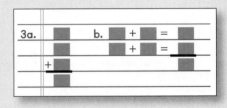

Lesson 89

Part 4 Write the dollars and cents number: $■■.■■

4a.	b.

a.

b.

Part 5 Write the answer to each problem.

a. $40 + 40 = $ ■   b. $120 - 20 = $ ■

c. $150 + 10 = $ ■   d. $60 - 30 = $ ■

5a.	b.
c.	d.

Part 6 Work each problem. Show the whole answer with a number and unit name.

a. There were 219 children in school. 120 were boys. How many were girls?

b. 18 of the children were sick. 89 of the children were well. How many children were there in all?

6a. →
b. →

Part 7 Write 4 facts for the family.

a.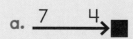

7a.	■ + ■ = ■
	■ + ■ = ■
	■ - ■ = ■
	■ - ■ = ■

Lesson 90

Part 1

1a.	b.	c.
d.	e.	f.

a.

b.

c.

d.

e.

f.

Part 2

a.   487
   + 97

b.   128
   − 62

c.   950
   − 117

d.   715
   − 194

Independent Work

Part 3 Work each problem. Show the whole answer with a number and a unit name.

a.   The red bike cost 27 dollars less than the green bike. The green bike cost 82 dollars. How many dollars did the red bike cost?

b.   Sam was 36 years younger than his mother. Sam was 9 years old. How old was his mother?

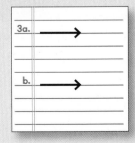

Connecting Math Concepts

Lesson 90

Part 4 Write the dollars and cents number: $\blacksquare\blacksquare.\blacksquare\blacksquare

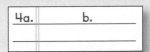

a.

b.

Part 5 Write and work the column problem for each item.

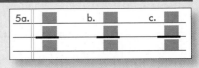

a. $38 + \blacksquare = 496$ b. $694 - \blacksquare = 256$ c. $\blacksquare + 509 = 878$

Part 6 Copy each problem and figure out the answer.

a. $\begin{array}{r} 593 \\ + \ 97 \end{array}$ b. $\begin{array}{r} 24 \\ 309 \\ +571 \end{array}$ c. $\begin{array}{r} 944 \\ -507 \end{array}$ d. $\begin{array}{r} 296 \\ -158 \end{array}$

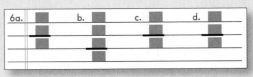

Part 7 Work each problem. Show the whole answer with a number and a unit name.

a. On Monday, the cheese store started out with 294 pounds of cheese. The store sold 185 pounds of cheese. How many pounds of cheese did the store still have?

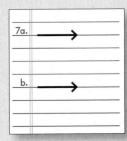

b. Ann started out with some quarters. Then her brother gave her 14 quarters. She ended up with 99 quarters. How many quarters did she start out with?

Lesson

Part 1 Find the perimeter of each figure.

10 in.

3 in.

a.

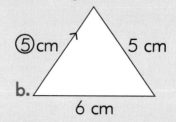
⑤cm   5 cm

b.

6 cm

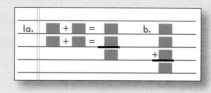

Part 2 Work each problem. Remember the unit name.

a. The red train was 190 feet longer than the green train. The red train was 449 feet long. How long was the green train?

b. The old boat was 17 feet longer than the new boat. The old boat was 56 feet long. How long was the new boat?

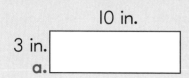

Part 3 Write the column problem for each item and work it.

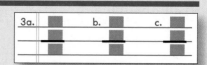

a. $32 + \blacksquare = 526$ b. $\blacksquare + 26 = 852$ c. $378 - \blacksquare = 167$

Part 4 Copy each problem and figure out the answer.

a. 34   b.   217   c.   572   d.   549
    64   + 193   − 380   − 79
  + 62

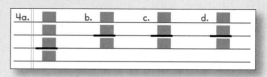

Part 5 Work each problem. Remember the unit name.

a. There are 85 boys and 79 girls at the play. How many children are at the play?

b. There were lots of animals on the farm. 69 of them were cows. The rest were pigs. There were 20 pigs on the farm. How many animals were on the farm?

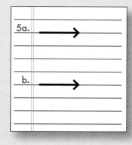

Connecting Math Concepts

Lesson 92

Independent Work

Part 1 Work each problem. Remember the unit name.

a. The truck started out with 711 boxes. The truck picked up more boxes. The truck ended up with 890 boxes. How many boxes did it pick up?

b. The train started out with 410 passengers. Then some of those passengers got off the train. The train ended up with 301 passengers. How many passengers got off the train?

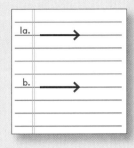

Part 2 Write the column problem for each item and work it.

a. $23 + \blacksquare = 141$ b. $696 - \blacksquare = 188$ c. $\blacksquare + 247 = 774$

Part 3 Copy each problem and figure out the answer.

a. 354
     66
   +502

b.   269
   + 745

c.   815
   − 506

d.   734
   − 690

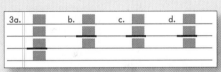

Part 4 Work each problem. Remember the unit name.

a. 66 people were on the bus. 53 of them were children. How many were adults?

b. Hank had 19 hammers and 16 saws. How many tools did he have?

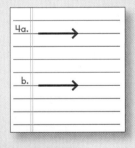

Lesson 93

Part 1 Write the column problem for each item and work it.

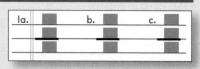

a. $26 + \blacksquare = 192$ b. $529 - \blacksquare = 466$ c. $\blacksquare + 334 = 928$

Part 2 Work each problem. Remember the unit name.

a. Mr. Brown was 88 centimers taller than his son. Mr. Brown was 179 centimeters tall. How tall was his son?

b. The blue bike cost 57 dollars more than the red bike. The blue bike cost 168 dollars. How much did the red bike cost?

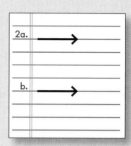

Part 3 Write the dollars and cents number: $\$\blacksquare\blacksquare.\blacksquare\blacksquare$

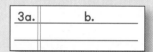

a.

b.

Part 4 Copy each problem and figure out the answer.

a. $\begin{array}{r} 526 \\ + 286 \end{array}$ b. $\begin{array}{r} 55 \\ 73 \\ + 13 \end{array}$ c. $\begin{array}{r} 650 \\ - 145 \end{array}$ d. $\begin{array}{r} 389 \\ - 298 \end{array}$

Part 5 Work each problem. Remember the unit name.

a. A bus started out with 87 children on it. Then 58 children got off the bus. How many children were still on the bus?

b. Mr. Briggs had 76 dollars. Then he spent 27 dollars in a store. How much money did he still have?

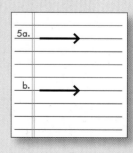

Lesson 94

Part 1 Write the time for each clock.

1a.	b.	c.	d.

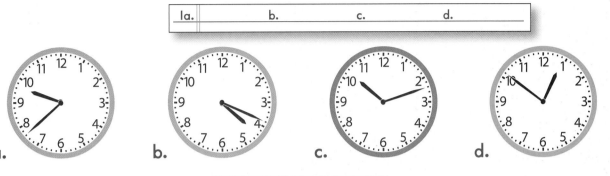

a.   b.   c.   d.

Independent Work

Part 2 Work each problem. Remember the unit name.

a.   590 children were in the zoo. 268 of the children were girls. How many boys were in the zoo?

b.   There were vehicles in a lot. 88 vehicles were cars. 82 vehicles were trucks. How many vehicles were in the lot?

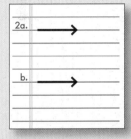

Lesson 94

Part 3 Find the perimeter of each figure.

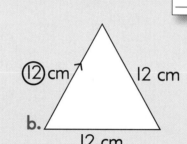

4 in.

10 in.

a.

b.

⑫ cm   12 cm

12 cm

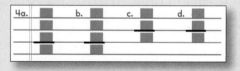

3a. ▪ + ▪ = ▪   b. ▪
▪ + ▪ = ▪
+ ▪

Part 4 Copy each problem and figure out the answer.

a. 104
    565
  + 279

b. 285
    535
  + 213

c. 654
  − 235

d. 649
  − 459

4a. ▪   b. ▪   c. ▪   d. ▪

Part 5 Work each problem. Remember the unit name.

a. Rob is 32 years younger than his mom. Rob is 9 years old. How many years old is his mom?

b. The brown fish was 13 inches shorter than the red fish. The brown fish was 16 inches long. How long was the red fish?

5a. →

b. →

Lesson 95

Part 1

a. $12 - 6 = $ ■
b. $7 + 6 = $ ■
c. $13 - 6 = $ ■
d. $14 - 8 = $ ■
e. $6 + 6 = $ ■

f. $13 - 7 = $ ■
g. $8 + 6 = $ ■
h. $14 - 6 = $ ■
i. $6 + 8 = $ ■
j. $6 + 7 = $ ■

1a.	f.
b.	g.
c.	h.
d.	i.
e.	j.

Part 2

a. $9 \times \underline{} = 72$
b. $2 \times \underline{} = 14$
c. $4 \times \underline{} = 20$
d. $10 \times \underline{} = 30$

2a.	b.
c.	d.

Part 3

3a.	b.	c.	d.

a.

b.

c.

d.

Part 4

a.

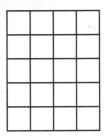

b.

4a.	
b.	

Lesson 95

Part 5 Write the estimation problem and the answer.
Remember the unit name.

a. There were 23 birds in John's yard. There were 46 birds in Mary's yard. About how many birds were in both yards?

5a.	
b.	

b. 38 birds are on a wire. 11 birds will fly away. About how many birds will still be on the wire?

Part 6 Write the column problem for each item and work it.

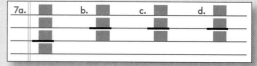

a. $46 + \blacksquare = 351$

b. $317 + \blacksquare = 752$

Part 7 Copy each problem and figure out the answer.

a.
$$65$$
$$46$$
$$+ 11$$

b.
$$435$$
$$+ 777$$

c.
$$692$$
$$- 345$$

d.
$$465$$
$$- 359$$

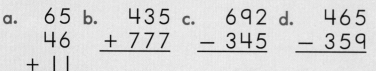

Connecting Math Concepts

Lesson 95

Part 8 Work each problem. Remember the unit name.

a. A train started out with 142 passengers. Then some passengers got off the train. The train ended up with 37 passengers. How many passengers got off the train?

b. There were some birds in a yard. After 6 more birds went to the yard, there were 71 birds in the yard. How many birds started out in the yard?

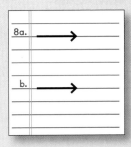

Part 9 Write the dollars and cents number: $■■.■■

  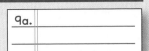

a.

Part 10

a.

b.

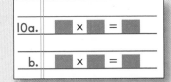

Part 11 Work each problem. Remember the unit name.

a. There were 87 trucks on a lot. 28 of the trucks were red. The others were black. How many trucks were black?

b. Tim ran 7 miles on Thursday. On Friday he ran 14 miles. How many miles did he run in all?

Lesson 96

Part 1

a. $14 - 6 = \blacksquare$

b. $7 + 6 = \blacksquare$

c. $6 + 8 = \blacksquare$

d. $13 - 6 = \blacksquare$

e. $13 - 7 = \blacksquare$

f. $8 + 6 = \blacksquare$

g. $14 - 8 = \blacksquare$

h. $6 + 7 = \blacksquare$

1a.	e.
b.	f.
c.	g.
d.	h.

Part 2

a. $5 \times \blacksquare = 45$

b. $9 \times \blacksquare = 45$

c. $2 \times \blacksquare = 12$

d. $10 \times \blacksquare = 40$

2a.	b.
c.	d.

Part 3

3a.	b.	c.	d.

a.

b.

c.

d.

Lesson 96

Part 4

4a.	
b.	

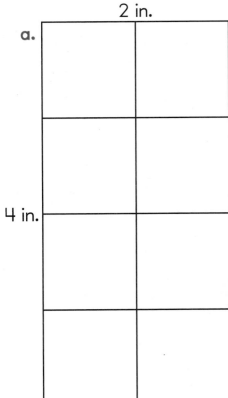

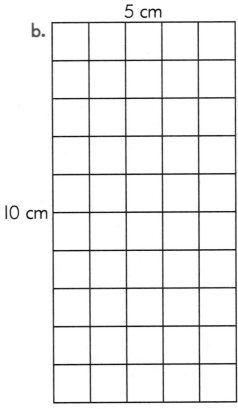

Part 5

a. $\begin{array}{r} \$8.90 \\ -6.15 \\ \hline \end{array}$

b. $\begin{array}{r} \$4.25 \\ +4.80 \\ \hline \end{array}$

c. $\begin{array}{r} \$1.48 \\ -\ .35 \\ \hline \end{array}$

d. $\begin{array}{r} \$6.88 \\ +1.28 \\ \hline \end{array}$

5a.	b.
c.	d.

Lesson

Independent Work

Part 6   Write the estimation problem and the whole answer.

a. It rained 22 days in May. It rained 9 days in June. About how many days did it rain in all?

b. Jim had 57 dollars. He gave 13 dollars to his mother. About how many dollars did he still have?

c. There were 12 children in the swimming pool. There were 19 children playing near the swimming pool. About how many children were there in all?

6a.	■ ■ = ■
b.	■ ■ = ■
c.	■ ■ = ■

Part 7   Write the column problem for each item and work it.

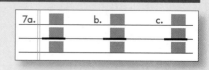

a. $417 + ■ = 822$ b. $839 - ■ = 211$ c. $■ + 617 = 841$

Part 8   Copy each problem and figure out the answer.

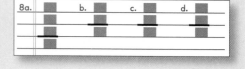

a.
```
  145
   27
+444
```
b.
```
  165
+  45
```
c.
```
  659
- 277
```
d.
```
  861
- 555
```

Lesson 96

Part 9 Find the perimeter of each figure.

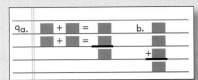

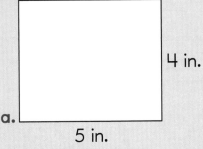

a.

4 in.

5 in.

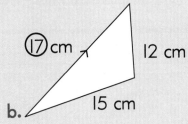

b.

⑰cm

12 cm

15 cm

Part 10

a.

b.

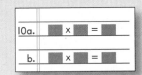

10a.	☐ x ☐ = ☐
b.	☐ x ☐ = ☐

Part 11 Work each problem.

a. Mary weighed 14 pounds more than her sister. Her sister weighed 57 pounds. How much did Mary weigh?

b. The store is 45 feet tall. The house is 60 feet tall. How many feet taller is the house than the store?

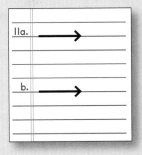

Lesson

Part 1

a. How many cents is 9 nickels?

b. How many cents is 6 nickels?

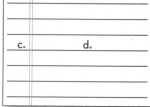

Part 2

2a.	b.	c.	d.

a.　　　　b.　　　　c.　　　　d.

Part 3

a. $ 4.75
　+ .25

b. $ 9.45
　−2.95

3a.	b.
c.	d.

c. $ 1.06
　− .52

d. $ 5.19
　+1.99

Part 4

4a.	
b.	
c.	

a.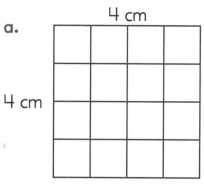

4 cm

4 cm

b.

5 cm

2 cm

c.

10 cm

1 cm

Lesson 97

Part 5 Copy and work each problem.

a. $4 \times \blacksquare = 32$ b. $2 \times \blacksquare = 18$

c. $5 \times \blacksquare = 20$ d. $9 \times \blacksquare = 27$

5a.	b.
c.	d.

Part 6 Work each problem.

a. 125 people went to the show. 19 adults went to the show. How many children went to the show?

b. Jan had 54 dollars. Fran had 47 dollars. How many dollars did the girls have in all?

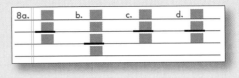

Part 7 Write the column problem for each item and work it.

a. $468 + \blacksquare = 918$ b. $968 - \blacksquare = 220$ c. $\blacksquare + 373 = 919$

Part 8 Copy each problem and figure out the answer.

a.
$$465$$
$$+444$$

b.
$$363$$
$$757$$
$$+\ \ \ 4$$

c.
$$928$$
$$-775$$

d.
$$829$$
$$-355$$

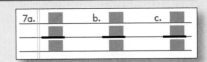

Part 9 Work each problem.

a. The blue boat was 56 feet longer than the yellow boat. The yellow boat was 99 feet long. How many feet long was the blue boat?

b. Chad weighs 78 pounds. Wally weighs 86 pounds. How many more pounds does Wally weigh than Chad?

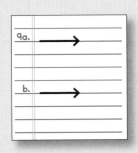

Lesson 98

Part 1

a. How many cents is 11 dimes?

b. How many cents is 3 quarters?

c. How many cents is 7 dimes?

d. How many cents is 4 nickels?

Part 2

a.

5 in.

3 in.

b.

6 cm

2 cm

c.

4 in.

1 in.

Connecting Math Concepts

Lesson 98

Part 3

a. $41.36
   +25.73

b. $82.50
   + 6.59

c. $136.12
   +   .50

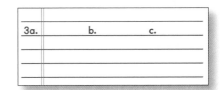
3a.   b.   c.

Independent Work

Part 4 Copy and work each problem.

a. $8.43
   +1.49

b.   628
   − 265

c.   28
     122
   + 43

d. $2.07
   − .95

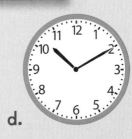
4a.   b.

c.   d.

Part 5 Write the time for each clock.

5a.   b.   c.   d.

a.

b.

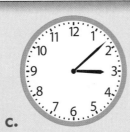
c.

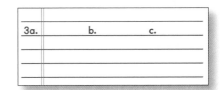
d.

Part 6 Write the estimation problem and the whole answer.

a. Sam had 88 dollars in one pocket. He had 23 dollars in another pocket. About how many dollars did Sam have?

b. Linda rode 12 miles on Monday and 17 miles on Tuesday. About how many miles did she ride on both days?

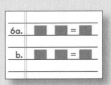
6a. ■ ■ = ■
b. ■ ■ = ■

Lesson 98

Part 7 Write the column problem for each item and work it.

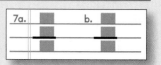

a. ■ + 473 = 819   b. 273 + ■ = 924

Part 8 Copy each problem and figure out the answer.

a.
$$267 + 45$$

b.
$$26$$
$$35$$
$$+31$$

c.
$$839 - 474$$

d.
$$917 - 762$$

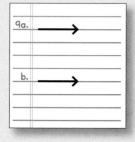

Part 9 Work each problem.

a. There are 29 children in the park. 18 of the children are boys. How many girls are in the park?

b. There were 28 clean socks in a drawer and 18 dirty socks in the laundry. How many socks were there in all?

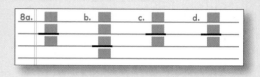

Connecting Math Concepts

Lesson 99

Part 1

a. How many cents is 4 quarters?

b. How many cents is 8 dimes?

c. How many cents is 7 nickels?

1a.	
b.	
c.	

Part 2

a.
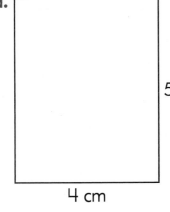
5 cm
4 cm

b.
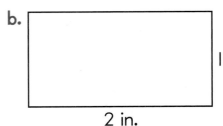
1 in.
2 in.

2a.	
b.	
c.	

c.

3 ft

10 ft

Part 3

a. $\begin{array}{r} \$89.47 \\ -5.29 \\ \hline \end{array}$

b. $\begin{array}{r} \$.93 \\ +120.25 \\ \hline \end{array}$

c. $\begin{array}{r} \$.26 \\ 4.06 \\ +.02 \\ \hline \end{array}$

3a.	b.	c.

Lesson 99

Part 4
Write the dollars and cents number: $■.■■

a.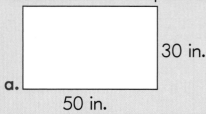

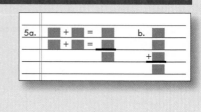

4a.

Part 5
Find the perimeter of each figure.

a.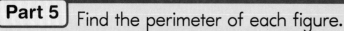
30 in.
50 in.

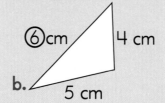
⑥cm 4 cm
b.
5 cm

5a. ■ + ■ = ■   b.
■ + ■ = ■
_____
■   + __

Part 6
Write the time for each clock.

a.

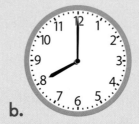
b.

6a.   b.

Part 7
Copy and work each problem.

7a.   b.   c.

a. $10 \times \blacksquare = 70$ b. $5 \times \blacksquare = 30$ c. $10 \times \blacksquare = 40$

Part 8
Copy each problem and work it. Remember the $ sign.

a. $2.63
  +4.36

b. $9.50
  −2.24

8a.   b.

Part 9
Write the estimation problem and the answer.

a. The baby had 8 toys. Then the baby got 11 more toys. About how many toys did the baby end up with?

b. There were 94 bananas in the store. The store sold 63 bananas. About how many bananas did the store end up with?

c. There were 38 girls in the park. Then 16 girls went home. About how many girls were still in the park?

9a. ■ ■ = ■
b. ■ ■ = ■
c. ■ ■ = ■

Lesson 100

Part 1

$9.50 $16.30 $24.99 $40.00

a. A person buys the book and the hat.

b. A person buys the shoes and the phone.

c. A person buys the hat, the shoes, and the phone.

Part 2

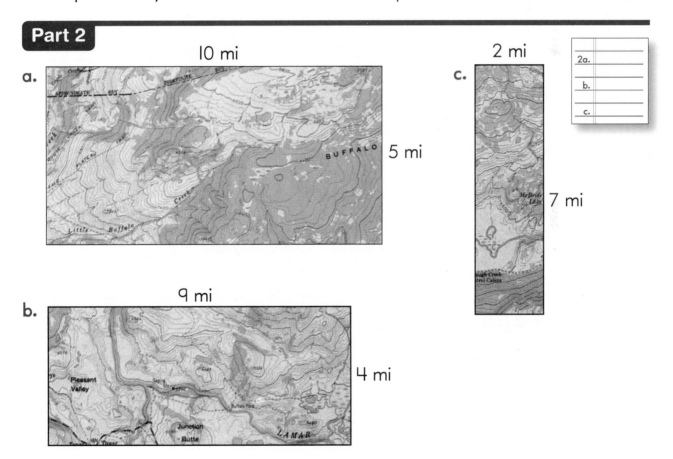

a. 10 mi

BUFFALO 5 mi

b. 9 mi

4 mi

c. 2 mi

7 mi

Lesson 100

Independent Work

Part 3

a.

b. How many cents is 8 nickels?

c. How many cents is 2 quarters?

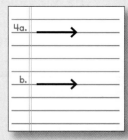

Part 4 Work each problem.

a. Dan had 11 fewer dollars than Bill had. Dan had 89 dollars. How many dollars did Bill have?

b. The red book costs $4.34. The blue book costs $9.67. How much more does the blue book cost than the red book?

Part 5 Write the column problem for each item and work it.

a. $\blacksquare + 354 = 827$ b. $578 - \blacksquare = 436$

Part 6 Copy each problem and figure out the answer.

a.
$$\begin{array}{r} 54 \\ + 46 \\ \hline \end{array}$$

b.
$$\begin{array}{r} 64 \\ + 355 \\ \hline \end{array}$$

c.
$$\begin{array}{r} 849 \\ - 575 \\ \hline \end{array}$$

d.
$$\begin{array}{r} 918 \\ - 363 \\ \hline \end{array}$$

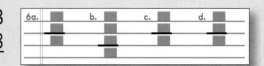

Part 7 Work each problem.

a. There were 28 children in the class. 14 of them were girls. How many were boys?

b. Ted put peanuts and cashew nuts in a bag. There are 39 nuts in the bag. 15 are peanuts. How many are cashew nuts?

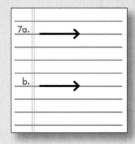

Connecting Math Concepts

Lesson 101

Part 1

| $2.20 | $12.50 | $2.99 | $7.05 |

a. A person buys the cup and the scarf.

b. A person buys the cup, the pencils, and the notebook.

c. A person buys the scarf and the pencils.

Part 2

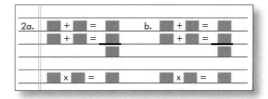

a.
5 ft

3 ft

b.

10 mi

9 mi

Part 3

a. $4 \times 6 = \blacksquare$

b. $9 \times \blacksquare = 63$

c. $2 \times \blacksquare = 10$

d. $5 \times 8 = \blacksquare$

e. $10 \times \blacksquare = 60$

f. $9 \times 4 = \blacksquare$

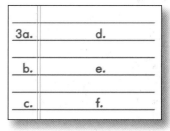

Lesson 101

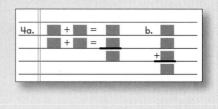

Independent Work

Part 4 Find the perimeter of each figure.

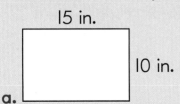

15 in.

10 in.

a.

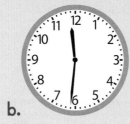

⑦cm

3 cm

5 cm

b.

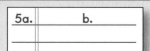

4a. ▮ + ▮ = ▮   b. ▮

     ▮ + ▮ = ▮

Part 5 Write the time for each clock.

5a.   b.

a.

b.

Part 6 Write 4 facts for each family.

a. 7   6 ➙ ■

b. 8   6 ➙ ■

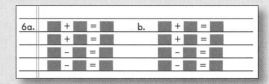

6a. ▮ + ▮ = ▮   b. ▮ + ▮ = ▮
    ▮ + ▮ = ▮   ▮ + ▮ = ▮
    ▮ − ▮ = ▮   ▮ − ▮ = ▮
    ▮ − ▮ = ▮   ▮ − ▮ = ▮

Part 7 Work each problem.

a. A truck started out with 460 bottles. Then the driver took 46 of those bottles off the truck. How many bottles were still on the truck?

b. In the morning, the train went 170 miles. In the afternoon, the train went 50 more miles. How far did the train end up going?

c. There were some children in the park. Then 77 more children came to the park. 194 children ended up in the park. How many children started out in the park?

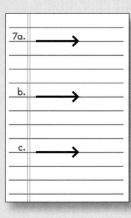

7a. ⟶

b. ⟶

c. ⟶

Part 8 Work a times problem for each item. Then write the unit name.

a. How many cents is 8 nickels?

b. How many cents is 4 dimes?

c. How many cents is 2 quarters?

8a.

b.

c

Lesson 102

Part 1

$31.60 $44.25 $12.95 $10.00

a. You want to buy the boots and the gloves. How much money do you need?

b. You want to buy the coat and the shirt. How much money do you need?

c. You want to buy the boots, the shirt, and the gloves. How much money do you need?

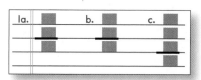

Part 2

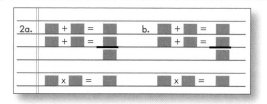

a.
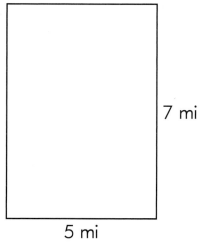

7 mi

5 mi

b.

4 ft

9 ft

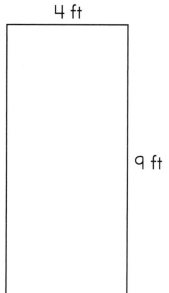

Lesson 102

Independent Work

Part 3 Copy and work each problem.

a. $5 \times \blacksquare = 15$

b. $10 \times \blacksquare = 40$

c. $9 \times 6 = \blacksquare$

d. $2 \times 8 = \blacksquare$

e. $2 \times \blacksquare = 12$

f. $4 \times 5 = \blacksquare$

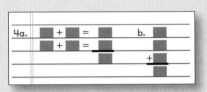

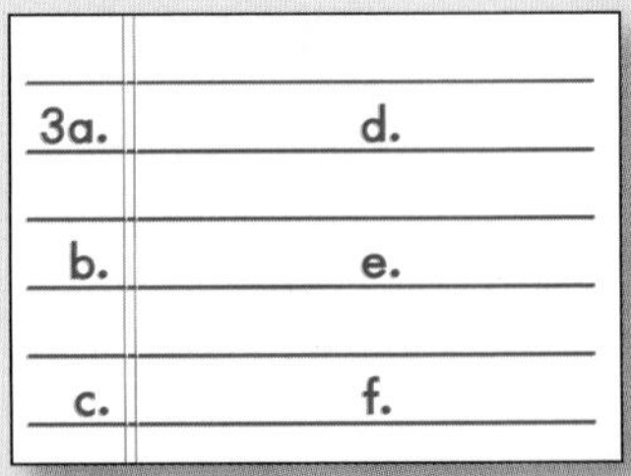

Part 4 Find the perimeter of each figure.

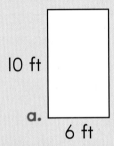

10 ft

a.

6 ft

b.

⑩ in.   10 in.

14 in.

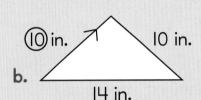

Part 5 Copy each problem and figure out the answer.

a. $\begin{array}{r} 734 \\ -444 \\ \hline \end{array}$

b. $\begin{array}{r} 524 \\ -272 \\ \hline \end{array}$

c. $\begin{array}{r} 458 \\ +534 \\ \hline \end{array}$

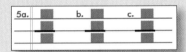

Part 6 Work the column problem for each item.

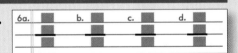

a. $\blacksquare + 47 = 74$

b. $6 + \blacksquare = 11$

c. $35 + \blacksquare = 140$

d. $29 - \blacksquare = 12$

Part 7 Work each problem.

a. There were 50 dogs in the park. 12 dogs were walking. The rest were running. How many dogs were running?

b. Mary read 7 books in May and 13 books in June. How many books did she read altogether?

c. There were 120 birds near the lake. 19 of them were red. The others were white. How many birds were white?

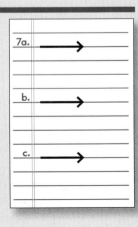

Connecting Math Concepts

Lesson 103

Part 1

a. Write 62 cents with a $ sign.

b. Write 80 cents with a $ sign.

c. Write 10 cents with a $ sign.

d. Write 12 cents with a $ sign.

1a.	
b.	
c.	
d.	

Part 2

$1.95 $5.50 $.45 $2.19

a. You want to buy the toothbrush and the hairbrush. How much money do you need?

b. You want to buy the hairbrush, the soap, and the toothpaste. How much money do you need?

c. You want to buy the toothbrush and the toothpaste. How much money do you need?

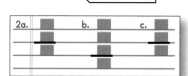

Independent Work

Part 3 Find the perimeter and area of each rectangle.

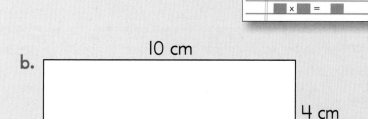

a.

2 in.

8 in.

b.

10 cm

4 cm

Lesson 103

Part 4 Copy and work each problem.

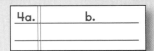

  a. $2 \times \blacksquare = 16$ b. $5 \times \blacksquare = 30$

Part 5 Copy each problem and figure out the answer.

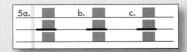

  a. $\begin{array}{r} 427 \\ -356 \end{array}$ b. $\begin{array}{r} 724 \\ -632 \end{array}$ c. $\begin{array}{r} 587 \\ +387 \end{array}$

Part 6 Write the time for each clock.

  a.   b.

Part 7 Write 2 subtraction facts for each family.

a. $\xrightarrow{\quad 7 \quad 6 \quad} \blacksquare$ b. $\xrightarrow{\quad 8 \quad 6 \quad} \blacksquare$

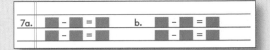

Part 8 Work each problem.

a. Dogs and cats were walking. There were 26 more dogs than cats. 50 dogs were walking. How many cats were walking?

b. There were 43 fewer blue dishes than white dishes. There were 27 blue dishes. How many white dishes were there?

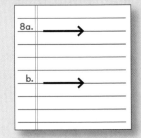

Part 9 Work a times problem for each item. Then write the unit name.

  a. How many cents is 5 nickels?

  b. How many cents is 8 dimes?

  c. How many cents is 9 nickels?

Lesson 104

Part 1

a. Write 400 cents with a $ sign.

b. Write 900 cents with a $ sign.

c. Write 80 cents with a $ sign.

d. Write 10 cents with a $ sign.

e. Write 9 cents with a $ sign.

f. Write 1 cent with a $ sign.

1a.	
b.	
c.	
d.	
e.	
f.	

Part 2

a. $11 - 3 = \blacksquare$

b. $4 + 8 = \blacksquare$

c. $12 - 8 = \blacksquare$

d. $13 - 5 = \blacksquare$

e. $5 + 8 = \blacksquare$

f. $12 - 4 = \blacksquare$

g. $11 - 8 = \blacksquare$

h. $13 - 8 = \blacksquare$

i. $8 + 5 = \blacksquare$

j. $8 + 4 = \blacksquare$

2a.	f.
b.	g.
c.	h.
d.	i.
e.	j.

Part 3

$1.95 $5.50 $.45 $2.19

$10.75

a. You buy the hairbrush.

b. You buy the toothpaste.

c. You buy the soap.

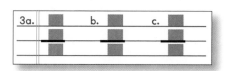

Lesson 104

Part 4 Copy and work each problem.

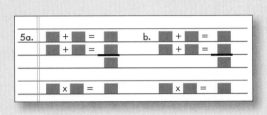

a. $10 \times \blacksquare = 70$ b. $5 \times \blacksquare = 20$

Part 5 Find the perimeter and area of each rectangle.

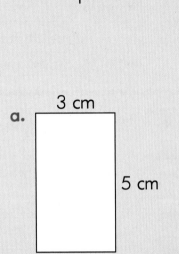

a.

3 cm

5 cm

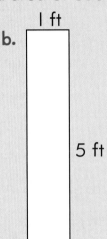

b.

I ft

5 ft

Part 6 Copy and work each problem.

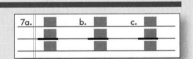

a. $\begin{array}{r} 3\,8\,1 \\ -\ 1\,7\,5 \end{array}$ b. $\begin{array}{r} 9\,1\,9 \\ -\ 6\,6\,6 \end{array}$ c. $\begin{array}{r} 4\,9\,2 \\ +\ 1\,8\,7 \end{array}$

Part 7 Work the column problem for each item.

a. $61 + \blacksquare = 195$ b. $\blacksquare + 25 = 161$ c. $270 - \blacksquare = 65$

Part 8 Write the time for each clock.

a.

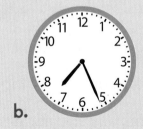

b.

Lesson 105

Part 1

a. Write 600 cents with a $ sign.

b. Write 605 cents with a $ sign.

c. Write 43 cents with a $ sign.

d. Write 4 cents with a $ sign.

e. Write 8 cents with a $ sign.

f. Write 50 cents with a $ sign.

1a.	
b.	
c.	
d.	
e.	
f.	

Part 2

a. $13 - 5 = \blacksquare$

b. $8 + 4 = \blacksquare$

c. $5 + 8 = \blacksquare$

d. $12 - 4 = \blacksquare$

e. $12 - 8 = \blacksquare$

f. $8 + 5 = \blacksquare$

g. $13 - 8 = \blacksquare$

h. $4 + 8 = \blacksquare$

2a.	e.
b.	f.
c.	g.
d.	h.

Part 3

 $9.50

 $16.30

 $24.65

 $40.00

$59.80

a. You buy the hat.

b. You buy the shoes.

c. You buy the phone.

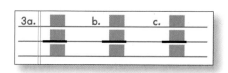

Lesson 105

Part 4 Copy and work each problem.

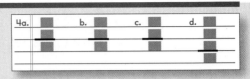

a. 570
− 239

b. 624
− 315

c. 408
− 193

d. 16
56
+71

Part 5 Figure out the missing number.

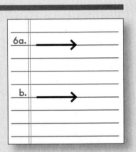

a. $\blacksquare + 4 = 11$ b. $3 + \blacksquare = 12$

c. $\blacksquare + 4 = 10$ d. $11 - \blacksquare = 5$

Part 6 Work each problem.

a. There were some children in the playground. Then 19 children left. There were still 84 children in the playground. How many children started out in the playground?

b. A train started out with 330 passengers. Then 26 passengers got off the train. How many passengers were still on the train?

Part 7 Write 4 facts for the family.

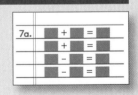

a.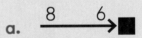

Part 8 Copy and work each problem.

a. $5 \times \blacksquare = 15$ b. $2 \times \blacksquare = 12$ c. $10 \times 7 = \blacksquare$

Lesson 105

Part 9 Find the perimeter and area of each rectangle.

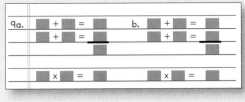

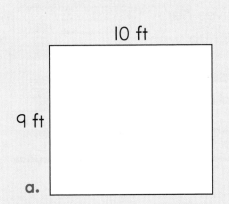

10 ft

9 ft

a.

5 in.

2 in.

b.

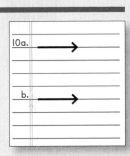

Part 10 Work each problem.

a. There were 55 children in the room. 36 were girls. How many were boys?

b. There were cats and dogs in the animal shelter. There were 125 cats. There were 55 dogs. How many animals were at the shelter?

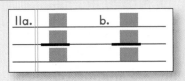

Part 11 Write the estimation problem and the answer.

a. About how much is 63 + 28?

b. About how much is 47 + 52?

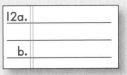

Part 12 Work a times problem for each item. Then write the unit name.

a. How many cents is 10 nickels?

b. How many cents is 10 dimes?

Lesson 106

Part 1

a. A woman has $3.12. Then her sister gives her $6.35. How much money does she end up with?

b. Dan starts out with $9.78. He spends some money. He ends up with $2.10. How much money does he spend?

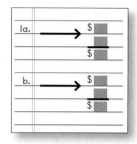

Part 2

$31.60 $12.95 $44.25 $10.00

$55.75

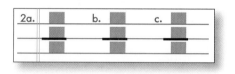

a. You buy the shirt. How much money do you still have?

b. You buy the boots. How much money do you still have?

c. You buy the coat. How much money do you still have?

Lesson 106

Part 3

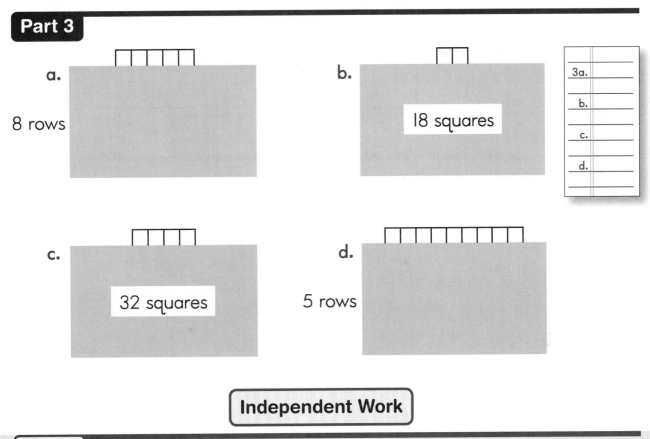

a.
8 rows

b.
18 squares

c.
32 squares

d.
5 rows

3a.	
b.	
c.	
d.	

Independent Work

Part 4

a.   Write 420 cents with a $ sign.

b.   Write 900 cents with a $ sign.

c.   Write 102 cents with a $ sign.

d.   Write 7 cents with a $ sign.

e.   Write 300 cents with a $ sign.

f.   Write 87 cents with a $ sign.

4a.	
b.	
c.	
d.	
e.	
f.	

Lesson 106

Part 5 Write 2 addition facts for each family.

a. $\xrightarrow{\quad 7 \qquad 6 \quad}$ ■

b. $\xrightarrow{\quad 8 \qquad 6 \quad}$ ■

Part 6 Work each problem.

a. Andy weighed 13 pounds more than Jim. Andy weighed 122 pounds. How many pounds did Jim weigh?

b. The red table weighed 27 pounds. The black table weighed 14 pounds. How many more pounds did the red table weigh than the black table?

c. The brown train was 321 feet shorter than the red train. The brown train was 605 feet long. How long was the red train?

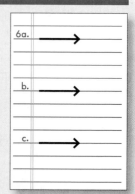

Lesson 106

Part 7 Write the column problem for each item and work it.

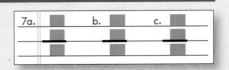

a. $\blacksquare + 18 = 71$ b. $86 + \blacksquare = 294$ c. $439 - \blacksquare = 165$

Part 8 Write the time for each clock.

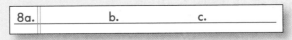

  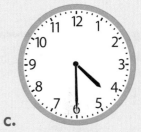

a. b. c.

Part 9 Write the dollars and cents number: $\$\blacksquare.\blacksquare\blacksquare$

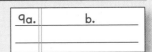

a.

b.

Part 10 Find the perimeter and the area of the rectangle.

7 cm

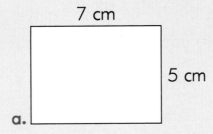

5 cm

a.

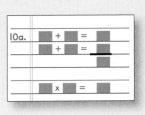

Lesson 107

Part 1

a. Jane has $8.82 more than Bob has. Bob has $.16. How much does Jane have?

b. Rob has $2.30 more than Mary. Rob has $7.28. How much money does Mary have?

c. Jan had $3.15 more than Pete had. Jan had $4.75. How much money did Pete have?

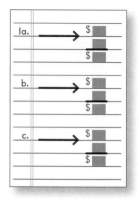

Part 2

 $31.60 $12.95 $44.25 $10.00

 $75.00

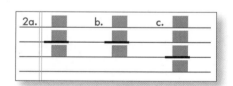

a. You want to buy the boots and the coat. Do you have enough money?

b. You want to buy the shirt and the coat. Do you have enough money?

c. You want to buy the boots, the shirt, and the gloves. Do you have enough money?

Lesson 107

Part 3

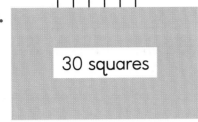
a.

30 squares

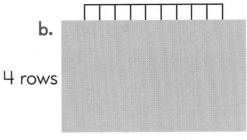
b.

4 rows

c.

3 rows

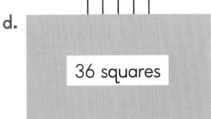
d.

36 squares

3a.	
b.	
c.	
d.	

Independent Work

Part 4 | Copy and work each problem.

4a.	b.	c.

a. 10 x 6 = ■   b. 5 x ■ = 20   c. 2 x ■ = 20

Part 5 | Write the estimation problem and the answer.

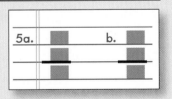

a. About how much is 89 + 70?

b. About how much is 43 + 19?

Part 6 | Copy and work each problem.

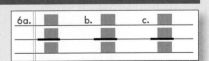

a. 319
  − 257

b. $4.34
  + 3.77

c. 218
  − 143

Lesson 107

Part 7 Write the column problem for each item and work it.

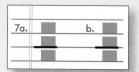
7a. ⬛ b. ⬛

a. $423 - \blacksquare = 107$ b. $38 + \blacksquare = 152$

Part 8 Work a times problem for each item. Then write the unit name.

a. How many cents is 4 dimes?

b. How many cents is 6 nickels?

8a.

b.

Part 9 Find the perimeter of each figure.

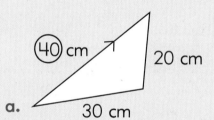

a. 40 cm   20 cm   30 cm

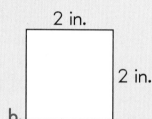
2 in.   2 in.

b.

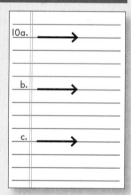
9a. ⬛   b. ⬛ + ⬛ = ⬛

Part 10 Work each problem.

a. There are 28 girls and 25 boys in the gym. How many children are in the gym?

b. There are 186 students in the lunchroom. 125 students are eating hot lunches. The rest are eating cold lunches. How many students are eating cold lunches?

c. 98 women and 85 men are at a wedding. How many people are at the wedding?

10a. →

b. →

c. →

Lesson 108

Part 1

a.
$$\begin{array}{r} 58 \\ +218 \\ \hline 286 \end{array}$$

b.
$$\begin{array}{r} 169 \\ +\ 28 \\ \hline 197 \end{array}$$

c.
$$\begin{array}{r} 145 \\ +235 \\ \hline 381 \end{array}$$

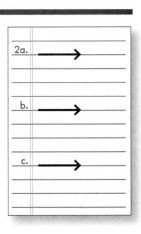

Part 2

a. Jill started out with $8.05. She spent $4.15. How much money did she end up with?

b. Pam had some money. She spent $3.49. She still had $2.45. How much money did she start with?

c. Rob had $3.88. Then his mom gave him some more money. He ended up with $9.28. How much money did his mom give him?

Part 3

$3.50

$1.99

$12.25

$3.20

$15.50

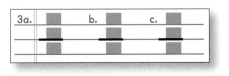

a. You want to buy the candles and the cake. Do you have enough money?

b. You want to buy the cake and the ice cream. Do you have enough money?

c. You want to buy the balloons and the cake. Do you have enough money?

Connecting Math Concepts

Lesson 108

Part 4

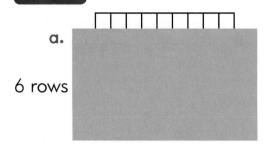

a.

6 rows

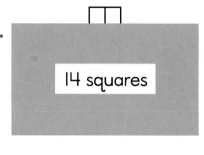

b.

14 squares

4a.
b.
c.
d.

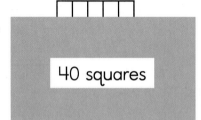

c.

40 squares

d.

10 rows

Independent Work

Part 5   Copy and work each problem.

a. $\begin{array}{r} 317 \\ 656 \\ +\ 111 \\ \hline \end{array}$ b. $\begin{array}{r} 473 \\ -367 \\ \hline \end{array}$ c. $\begin{array}{r} 649 \\ -283 \\ \hline \end{array}$ d. $\begin{array}{r} 634 \\ +766 \\ \hline \end{array}$

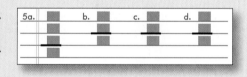

Part 6   Work each estimation problem.

a.   Mary ran 52 miles last month and 49 miles this month. About how many miles did she run in all?

b.   There were 77 kids in the park. 48 kids went home. About how many kids are still in the park?

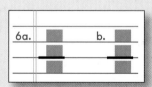

Lesson 108

Part 7

Write the dollars and cents number: $■■.■■

7a.	b.

a.

b.

Part 8

Find the perimeter and the area of the rectangle.

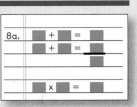

12 in.

a. [   ] 1 in.

Part 9

Write the time for each clock.

9a.	b.	c.

a.

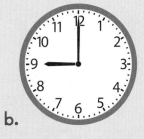
b.

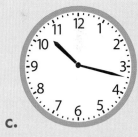
c.

Part 10

Copy and work each problem.

10a.	b.	c.

a. 2 x ■ = 14   b. 5 x 8 = ■   c. 10 x ■ = 40

Lesson 108

Part 11 Work each problem.

a. There were red cars and blue cars in the lot. There were 16 more red cars than blue cars. There were 33 blue cars. How many red cars were there?

b. Tammy was 27 years older than her son Bill. Tammy was 82 years old. How old was her son?

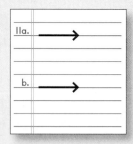

Part 12 Copy and work each problem.

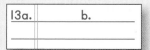

a.
```
  587
- 269
```

b.
```
  $4.51
+ 3.67
```

c.
```
  $12.07
+ 13.79
```

Part 13

a. Write 405 cents with a $ sign.

b. Write 763 cents with a $ sign.

Lesson 109

Part 1

a. Liz had $9.76. Henry had $2.81. How much more money did Liz have than Henry had?

b. Jerry had $6.88 less than Bob. Jerry had $2.90. How much money did Bob have?

1a.	→
b.	→

Part 2

	cars	buses	trucks
street	20	4	15
mall	35	10	22
park	14	6	8

a. How many trucks are at the mall?

b. How many buses are in the park?

c. How many cars are in the mall?

d. How many trucks are in the street?

2a.	
b.	
c.	
d.	

Part 3

a.
```
  1 2 3
+   1 4
-------
  1 3 8
```

b.
```
  4 0 8
+ 1 5 2
-------
  5 6 0
```

c.
```
    5 8
+ 2 1 8
-------
  2 6 6
```

3a.	b.
c.	

Part 4

 $3.50

 $1.99

 $12.25

 $3.20

 $14.90

a. You buy the balloons and the ice cream.

b. You buy the candles and the ice cream.

c. You buy the cake and the candles.

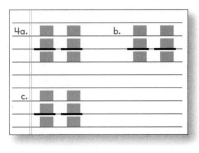

Connecting Math Concepts

Lesson 109 **139**

Lesson 109

Part 5 Find the rows or the squares.

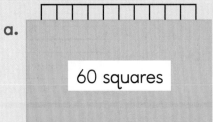

a.

60 squares

b.

7 rows

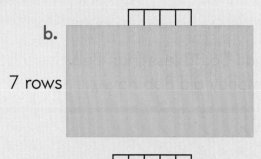

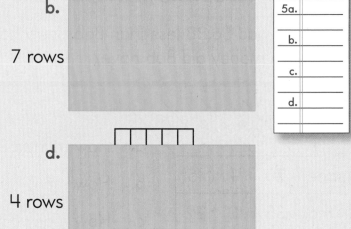

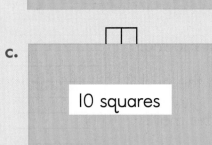

c.

10 squares

d.

4 rows

Part 6 Work the column problem for each item.

a. ■ + 668 = 894   b. ■ − 876 = 166

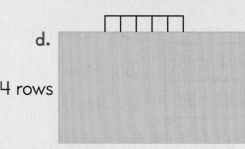

Part 7

a. Write 500 cents with a $ sign.

b. Write 283 cents with a $ sign.

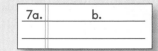

Part 8 Copy and work each problem.

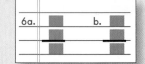

a. 5 x ■ = 30   b. 2 x 5 = ■   c. 10 x ■ = 10

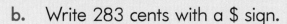

Part 9 Write the estimation problem and the answer.

a. About how much is 134 + 36?

b. About how much is 239 + 19?

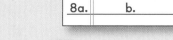

Part 10 Work a times problem for each item. Then write the unit name.

a. How many cents is 7 dimes?

b. How many cents is 6 nickels?

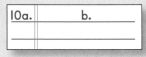
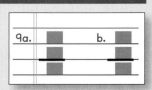
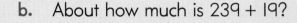

Connecting Math Concepts

Lesson 109

Part 11 Copy and work each problem.

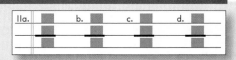

a. $\begin{array}{r} 25 \\ +487 \\ \hline \end{array}$

b. $\begin{array}{r} 472 \\ +465 \\ \hline \end{array}$

c. $\begin{array}{r} 473 \\ -356 \\ \hline \end{array}$

d. $\begin{array}{r} 16 \\ +71 \\ \hline \end{array}$

Part 12 Write the time for each clock.

| 12a. | b. | c. |

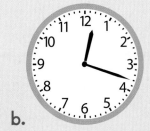

a.   b.   c.

Part 13 Find the perimeter and the area of each rectangle.

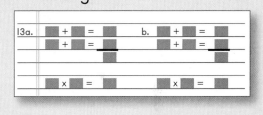

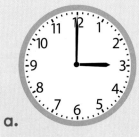

1 ft

6 ft

a.

b.

10 in.

6 in.

Part 14 Work each problem.

a. A truck started out with 120 sheep. The truck picked up more sheep. The truck ended up with 210 sheep. How many sheep did the truck pick up?

b. Dave had some rocks. He sold 51 rocks and ended up with 16 rocks. How many rocks did he start with?

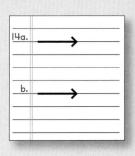

Lesson 110

Part 1

R, T, C, S, H

1.	2.	3.	4.
5.	6.	7.	8.

1.

2.

3.

4.

5.

6.

7.

8.

Part 2

  $2.20

  $12.50

  $2.99

 $7.05

 $28.80

a.   You buy the cup and the pencils.

b.   You buy the cup, the scarf, and the notebook.

Part 3

	black	green	blue
skirts	72	16	31
blouses	21	45	4
pants	12	7	27

a.   How many blue pants are there?

b.   How many black skirts are there?

c.   How many green blouses are there?

d.   How many skirts are green?

e.   How many pants are black?

3a.
b.
c.
d.
e.

Lesson 110

Part 4

a.
```
  718
+ 242
  960
```

b.
```
  918
+  42
  961
```

c.
```
  417
+ 153
  560
```

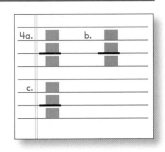

Independent Work

Part 5 Work each problem.

a. Ronda has $2.30 more than Tina. Ronda has $43.88. How much does Tina have?

b. Susan has $35.24 less than Jerry. Susan has $85.06. How much money does Jerry have?

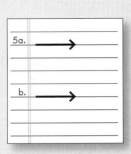

Part 6 Copy each problem and figure out the answer.

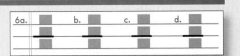

a.
```
  483
+ 168
```

b.
```
  973
- 566
```

c.
```
  724
- 706
```

d.
```
  77
+ 66
```

Part 7 Write the dollars and cents number: $■■.■■

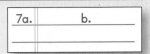

a.

b.

Part 8 Write the column problem for each item and work it.

a. ■ + 18 = 53

b. 27 + ■ = 74

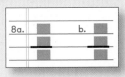

Lesson 110

Part 9 Write the fact for the missing number in each family.

a. 8 ■ → 13   b. 7 ■ → 12   c. 7 6 → ■

9a.	b.	c.
d.	e.	f.
g.		

d. 9 ■ → 15   e. 8 6 → ■   f. ■ 4 → 12

g. 5 ■ → 14

Part 10 Find the rows or the squares.

a. 30 squares   b. 36 squares

10a.	
b.	
c.	
d.	

c. 70 squares   d. 6 rows

Part 11 Work a times problem for each item. Then write the unit name.

a. How many cents is 5 quarters?

b. How many cents is 12 dimes?

11a.	
b.	

Lesson 111

Part 1

| $9.25 | $25.50 | $1.50 | $3.49 |

$26.75

1a.	b.
c.	

a. You want to buy the phone and the pencils.

b. You want to buy the book and the paper.

c. You want to buy the phone and the paper.

Part 2

	Bob	Kay	Mike
blue	3	6	1
white	4	3	5
green	0	5	2
red	1	6	0

Shirts

a. Who has 5 white shirts?

b. Who has the most red shirts?

c. Who has 0 green shirts?

d. How many blue shirts does Kay have?

e. How many green shirts does Mike have?

2a.	
b.	
c.	
d.	
e.	

Part 3

a.
$$\begin{array}{r} 86 \\ -39 \\ \hline 45 \end{array}$$

b.
$$\begin{array}{r} 75 \\ -19 \\ \hline 56 \end{array}$$

c.
$$\begin{array}{r} 425 \\ -210 \\ \hline 235 \end{array}$$

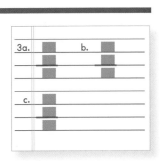

Lesson 111

Part 4

a.
27 apples

b.
5 rows

4a.	
b.	
c.	
d.	

c.
15 apples

d.
9 rows

Independent Work

Part 5 Write the time for each clock.

5a.	b.	c.

a.

b.

c.

Lesson 111

Part 6 Write the fact for the missing number in each family.

a. $\blacksquare \xrightarrow{\quad 5 \quad} 13$

b. $\blacksquare \xrightarrow{\quad 7 \quad} 13$

c. $8 \;\blacksquare \xrightarrow{\quad} 14$

d. $9 \;\blacksquare \xrightarrow{\quad} 14$

e. $\blacksquare \xrightarrow{\quad 6 \quad} 14$

f. $\blacksquare \xrightarrow{\quad 8 \quad} 17$

g. $7 \;\blacksquare \xrightarrow{\quad} 14$

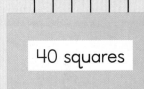

6a.	b.	c.
d.	e.	f.
g.		

Part 7 Find the rows or the squares.

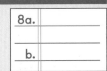

7a.	b.	c.

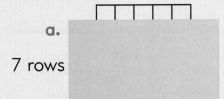

a.

7 rows

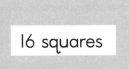

b.

16 squares

c.

40 squares

Part 8

a. How many cents is 8 quarters?

b. How many cents is 9 nickels?

8a.
b.

Part 9 Copy each problem and figure out the answer.

9a.	b.	c.

a.
$$\begin{array}{r} 53 \\ 17 \\ +95 \\ \hline \end{array}$$

b.
$$\begin{array}{r} 436 \\ +187 \\ \hline \end{array}$$

c.
$$\begin{array}{r} 836 \\ -174 \\ \hline \end{array}$$

Part 10 Work each problem.

a. Emily had $57.00 in bills and $31.29 in coins. How much money did she have altogether?

b. There was $86.48 in a bank. That money belonged to Jan and Dan. $25.84 belonged to Dan. How much belonged to Jan?

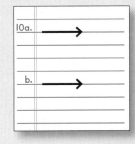

Part 11 Write the estimation problem and the answer.

a. About how much is 28 + 12?

b. About how much is 58 + 39?

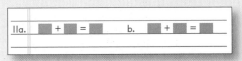

Lesson 112

Part 1

	oak	ash	pine
5 years	5 ft	6 ft	11 ft
10 years	9 ft	14 ft	24 ft
20 years	16 ft	26 ft	40 ft

a. How tall is the oak tree that is 20 years old?

b. How old is the pine tree that is 24 feet tall?

c. How tall is the ash tree that is 5 years old?

d. How old is the ash tree that is 26 feet tall?

1a.	
b.	
c.	
d.	

Part 2

a.
$$\begin{array}{r} 48 \\ + 43 \\ \hline 90 \end{array}$$

b.
$$\begin{array}{r} 95 \\ - 16 \\ \hline 81 \end{array}$$

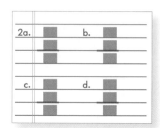

c.
$$\begin{array}{r} 170 \\ - 22 \\ \hline 148 \end{array}$$

d.
$$\begin{array}{r} 410 \\ + 156 \\ \hline 576 \end{array}$$

Part 3

a.

24 bottles

b.

45 bottles

3a.	
b.	
c.	
d.	

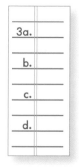

c.

45 bottles

d.

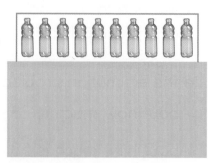

3 rows

Connecting Math Concepts

Lesson 112

Part 4 If you can't buy the items, write **no**. If you can buy the items, figure out how much money you still have.

 $2.20

 $2.00

$16.00

 $17.59

$20.60

4a.	b.
c.	

a.   You want to buy the toothpaste and the hammer.

b.   You want to buy the wallet, the socks, and the toothpaste.

c.   You want to buy the wallet and the toothpaste.

Part 5 Write the time for each clock.

5a.	b.

a.

b.

Part 6 Write the fact for the missing number in each family.

a. 8 6 → ■

b. 9 ■ → 15

c. 7 ■ → 12

6a.	b.	c.
d.	e.	f.
g.	h.	i.

d. 6 ■ → 13

e. 7 5 → ■

f. ■ 8 → 14

g. ■ 4 → 12

h. 6 ■ → 12

i. 8 ■ → 11

Connecting Math Concepts

Lesson 112

Part 7

   a.   How many cents is 11 nickels?

   b.   How many cents is 5 quarters?

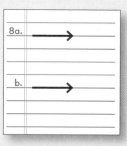

Part 8 Work each problem.

   a.   Vern started out with $14.73. Then he earned some more money. He ended up with $57.45. How much did he earn?

   b.   Mrs. Jones started out with $62.38. She spent some money and ended up with $20.75. How much did she spend?

Part 9 Write the estimation problem and the answer.

   a.   About how much is 27 + 44?

   b.   About how much is 24 + 79?

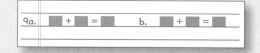

Lesson 113

Part 1

	oak	ash	pine
5 years	5 ft	6 ft	11 ft
10 years	9 ft	14 ft	24 ft
20 years	16 ft	26 ft	40 ft

a. How tall is the oak tree that is 5 years old?

b. How old is the pine tree that is 24 feet tall?

c. How old is the ash tree that is 6 feet tall?

d. How tall is the oak tree that is 20 years old?

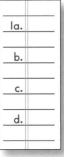

Part 2

a.
```
  128
+  58
  186
```

b.
```
   98
 - 59
   40
```

c.
```
   66
+ 125
  190
```

d.
```
  140
 - 12
  128
```

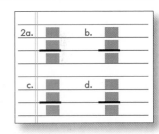

Part 3

a. How many girls are there on 5 teams? 35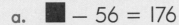

b. How many buses hold 100 people? 4 ▮

c. How many boxes hold 36 dolls? 9 ▮

d. How many slices are there in 6 pizzas? 48 ▮

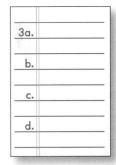

Independent Work

Part 4 Write the column problem for each item and work it.

a. ▮ − 56 = 176   b. 532 + ▮ = 871

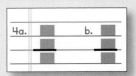

Part 5 Copy and work each problem.

a.
```
  484
- 368
```

b.
```
  833
- 327
```

c.
```
  549
- 176
```

Lesson 113

Part 6 — Write the fact for the missing number in each family.

a. $\underset{\xrightarrow{\hspace{1.2cm}}}{\underline{6 \quad \blacksquare}}\,15$

b. $\underset{\xrightarrow{\hspace{1.2cm}}}{\underline{8 \quad 6}}\,\blacksquare$

c. $\underset{\xrightarrow{\hspace{1.2cm}}}{\underline{3 \quad \blacksquare}}\,12$

d. $\underset{\xrightarrow{\hspace{1.2cm}}}{\underline{7 \quad 5}}\,\blacksquare$

e. $\underset{\xrightarrow{\hspace{1.2cm}}}{\underline{6 \quad \blacksquare}}\,13$

f. $\underset{\xrightarrow{\hspace{1.2cm}}}{\underline{9 \quad \blacksquare}}\,15$

g. $\underset{\xrightarrow{\hspace{1.2cm}}}{\underline{6 \quad \blacksquare}}\,11$

h. $\underset{\xrightarrow{\hspace{1.2cm}}}{\underline{6 \quad \blacksquare}}\,12$

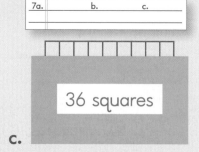

Part 7 — Find the rows.

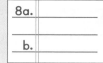

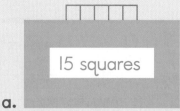

a. 15 squares

b. 16 squares

c. 36 squares

Part 8

a. How many cents is 3 dimes?

b. How many cents is 7 nickels?

Part 9 — Work each problem.

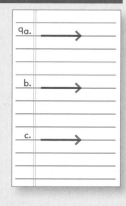

a. The white shoes cost $28.30 less than the brown shoes. The white shoes cost $26.98. How much did the brown shoes cost?

b. The brown gloves cost $12.88. The white gloves cost $31.59 more than the brown gloves. How much did the white gloves cost?

c. Don had $40.54 less than his brother. Don had $98.25. How much did his brother have?

Part 10 — Write the estimation problem and the answer.

a. About how much is 37 + 42?

b. About how much is 38 + 19?

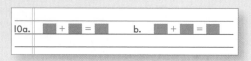

Connecting Math Concepts

Lesson 114

Part 1

	oak	ash	pine
5 years	5 ft	6 ft	11 ft
10 years	9 ft	14 ft	24 ft
20 years	16 ft	26 ft	40 ft

a. Which 5-year-old tree is the tallest?

b. Which 10-year-old tree is the shortest?

c. Which 20-year-old tree is 26 feet tall?

d. Which 10-year-old tree is 14 feet tall?

e. How old is the oak tree that is 16 feet tall?

f. How tall is the oak tree that is 10 years old?

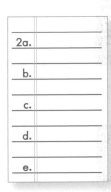

Part 2

a. How many children are in 8 boats?   80 ▬▬▬

b. How many boats hold 50 children?   5 ▬▬▬

c. How many bags hold 6 hot dogs?   3 ▬▬▬

d. How many apples are in 4 boxes?   36 ▬▬▬

e. How many boxes hold 32 cans?   4 ▬▬▬

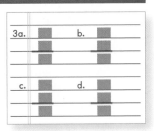

Part 3

a.
$$175 - 80 = 115$$

b.
$$119 + 143 = 262$$

c.
$$210 + 190 = 300$$

d.
$$350 - 260 = 90$$

Independent Work

Part 4   Write the time for each clock.

a.

b.

Lesson 114

Part 5 | Write the column problem for each problem and work it.

a. ■ + 88 = 194   b. 185 + ■ = 439

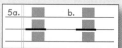

Part 6 | Write the fact for the missing number in each family.

a. ■ —5→ 13   b. 8 ■—→ 14   c. ■ —6→ 13

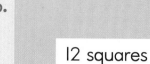

d. 8 ■—→ 11   e. ■ 8—→ 16   f. 6 ■—→ 12

g. ■ 4—→ 12   h. ■ 6—→ 14

Part 7 | Find the rows or the squares.

a.

14 squares

b.

12 squares

c.

3 rows

d.

40 squares

Part 8 | Copy and work each problem.

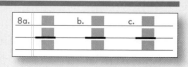

a.   285
   − 249

b.   346
   − 85

c.   266
   + 75

Part 9 | Write the estimation problem and the answer.

a.   About how much is 22 + 48?

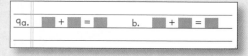

b.   About how much is 17 + 11?

Lesson 114

Part 10 If you can't buy the items, write **no.** If you can buy the items, figure out how much money you still have.

 $47.59

 $36.00

 $12.20

 $3.39

 $50.75

10a.	b.
c.	

a. You want to buy the video game and the notebook.

b. You want to buy the watch and the pens.

c. You want to buy the pens and the notebook.

Lesson 115

Part 1

a. How many rooms hold 100 people?   4 ▬▬▬▬

b. How many people fit around 8 tables?   48 ▬▬▬▬

c. How many marbles are in 3 bags?   150 ▬▬▬▬

d. How many cartons hold 144 eggs?   12 ▬▬▬▬

e. How many balloons are in 5 bags?   75 ▬▬▬▬

1a.	
b.	
c.	
d.	
e.	

Part 2

	Joe	Pam	Mary	Sid
1 year	$250	$300	$45	$0
3 years	$610	$650	$400	$75
5 years	$2000	$1550	$1200	$0

2a.	
b.	
c.	
d.	
e.	
f.	

a. Who saved $650 in 3 years?

b. How many years did it take Mary to save $1200?

c. How much did Joe save after 3 years?

d. Who saved the most in 1 year?

e. Who saved the least after 1 year?

f. How much did Sid save after 1 year?

Independent Work

Part 3   Check each answer. If an answer is wrong, find the correct answer.

3a.	b.
c.	

a.
$$
\begin{array}{r} 363 \\ +192 \\ \hline 555 \end{array}
$$

b.
$$
\begin{array}{r} 265 \\ -136 \\ \hline 139 \end{array}
$$

c.
$$
\begin{array}{r} 522 \\ +686 \\ \hline 1208 \end{array}
$$

Lesson 115

Part 4 Find the perimeter of each figure.

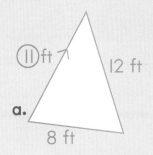

a.

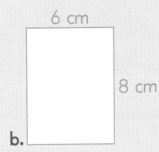

6 cm

8 cm

b.

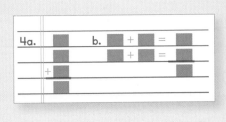

Part 5 Find the rows or the squares.

6 rows

a.

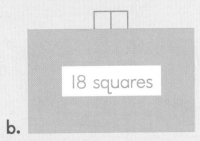

18 squares

b.

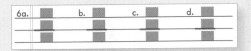

Part 6 Write the column problem for each item and work it.

a. ■ − 10 = 3   b. 17 + ■ = 24

c. 38 − ■ = 11   d. ■ + 16 = 30

Part 7 Work each problem.

a. Mary had 3 nickels and 15 dimes. How many coins did she have in all?

b. There were 165 children in the park. 82 of them were girls. How many were boys?

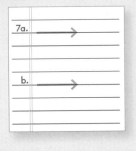

Part 8 Write the dollars and cents number: $■■.■■

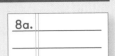

a.

Lesson 115

Part 9 Copy each problem and figure out the answer.

a. $ \$1.36 $
$ .22 $
$ +4.51 $

b. $ \$4.83 $
$ -3.26 $

c. $ \$5.03 $
$ +12.96 $

Part 10 Write the time for each clock.

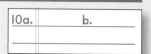

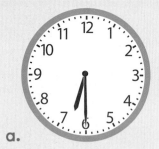

a.

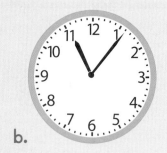

b.

Lesson 116

Independent Work

Part 1 The table shows how much money people saved.

	Joe	Pam	Mary	Sid
1 year	$250	$300	$45	$0
3 years	$610	$650	$400	$70
5 years	$2000	$1550	$1200	$0

1a.	b.
c.	d.

a. How much money did Joe save after 3 years?

b. Who saved the most money after one year?

c. How much money did Mary save after 5 years?

d. Who saved the least after 3 years?

Part 2 If you can't buy the items, write **no.** If you can buy the items, figure out how much money you still have.

$36.00 $19.26 $24.90 $14.50

$55.99

2a.	b.
c.	

a. You want to buy the shoes, the shirt, and the belt.

b. You want to buy the hat and the belt.

c. You want to buy the shoes and the hat.

Lesson 116

Part 3 Write the dollars and cents number: $■■.■■

3a. [_____]

a.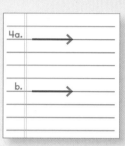

Part 4 Work each problem.

a. A train car had 963 boxes in it. Then workers took 536 boxes out of the train car. How many boxes were still in the train car?

b. Joanne weighed 134 pounds. Then she lost 28 pounds. How much does she weigh now?

4a. ⟶

b. ⟶

Part 5 Copy each problem and figure out the answer.

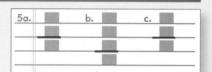

a. $ 5.76
   + .47

b. $ .23
    1.34
   + .30

c. $ 5.63
   −4.56

Part 6

6a. [b.   c.   d.   ]

a. Write 306 cents with a $ sign.

b. Write 170 cents with a $ sign.

c. Write 1000 cents with a $ sign.

d. Write 48 cents with a $ sign.

Part 7

a. How many cents is 5 quarters?

b. How many cents is 12 nickels?

7a. [_____]

b. [_____]

Connecting Math Concepts

Lesson 117

Part 1 Copy and work each problem.

a. $\begin{array}{r} 426 \\ +386 \end{array}$ b. $\begin{array}{r} 583 \\ -177 \end{array}$ c. $\begin{array}{r} 448 \\ -439 \end{array}$

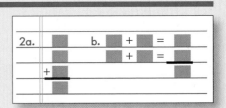

Part 2 Find the perimeter of each figure.

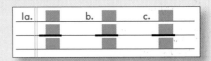

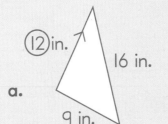

a. ⑫ in. 16 in. 9 in.

b. 10 ft 10 ft

Part 3 Find the rows or the squares.

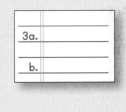

3a.

b.

a.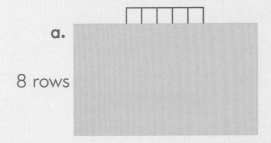

8 rows

b. 70 squares

Part 4 Write the dollars and cents number: $■■.■■

4a.

a.

Part 5 Write the column problem for each item and work it.

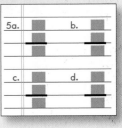

a. ■ + 41 = 70   b. 32 + ■ = 86

c. ■ − 55 = 90   d. 14 + ■ = 72

Lesson 117

Part 6 Write the time for each clock.

6a.	b.

a.

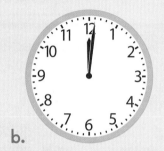

b.

Part 7

	small	medium	large
red	15	22	0
green	8	18	15
blue	21	11	30
white	19	14	6

Shirts

a. How many large white shirts are there?

b. The most green shirts are what size?

c. The most large shirts are what color?

d. How many red shirts are medium?

e. 21 of the blue shirts are what size?

7a.	
b.	
c.	
d.	
e.	

Lesson  118

Part 1 Write the place-value column problem for each number.

a. 4203   b. 14   c. 658

1a.	b.	c.

Part 2 Work each problem.

a. Jan had $5.30 more than Don had. Don had $24.25. How much did Jan have?

b. Alex weighed 31 pounds less than Greg. Greg weighed 90 pounds. How much did Alex weigh?

c. The red barn had 339 mice in it. The green barn had 199 mice in it. How many more mice were in the red barn than the green barn?

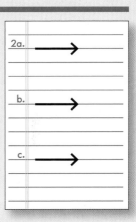

Part 3 Write the time for each clock.

3a.	b.

a.

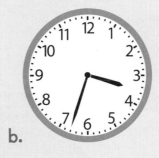

b.

Part 4 Copy and work each problem.

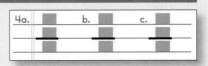

a. $5.26
  +6.97

b. $15.07
  + 1.97

c. $7.75
  −3.28

Lesson 118

Part 5

	Mary	Pat	Bob	Lane
March	$25	$20	$0	$35
April	$10	$40	$12	$50
May	$29	$16	$38	$15

5a.	
b.	
c.	
d.	
e.	

a. Who had the most money in April?

b. How much did Bob have in May?

c. Who had $10 in April?

d. In March, how much did Pat have?

e. Who had the least money in May?

Lesson 119

Part 1

a.

b.

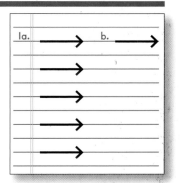

Part 2

a. How many cents is 6 quarters?

b. How many cents is 10 dimes?

2a.
b.

Part 3 Find the rows or the squares.

a.

11 rows

b.

44 squares

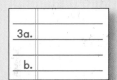

3a.
b.

Part 4 Write the column problem for each item and work it.

a. $68 + \blacksquare = 96$

b. $124 - \blacksquare = 79$

c. $\blacksquare - 34 = 18$

d. $\blacksquare + 33 = 91$

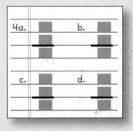

Part 5 Write the dollars and cents number: $\blacksquare\blacksquare.\blacksquare\blacksquare$

5a.

a.

Lesson 119

Part 6

Write the time for each clock.

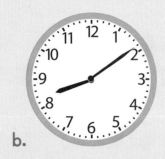

a.

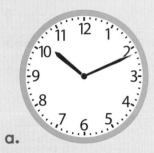

b.

6a.		b.

Part 7

Copy each problem and figure out the answer.

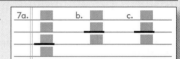

a.
$$\begin{array}{r} \$4.20 \\ 3.08 \\ +2.69 \\ \hline \end{array}$$

b.
$$\begin{array}{r} \$7.46 \\ -5.38 \\ \hline \end{array}$$

c.
$$\begin{array}{r} \$3.87 \\ +5.97 \\ \hline \end{array}$$

Part 8

Work each problem.

a. The farmer had 17 brown cows and some white cows. The farmer had 124 cows in all. How many were white cows?

b. Ben had 35 quarters, 3 nickels, and 16 pennies. How many coins did he have altogether?

Lesson 120

Part 1

a.

b.

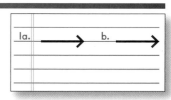
1a. ————→ b. ————→

Independent Work

Part 2

Write the dollars and cents number: $■■.■■

2a.

a.

Part 3

Copy and work each problem.

a. 263
     67
  +644

b. 581
  +798

c. 982
  −376

Part 4

Work each problem.

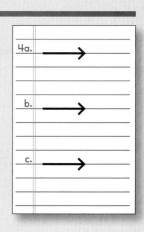
4a. ————→
b. ————→
c. ————→

a. A truck started out with some boxes. Then the truck dropped off 157 boxes. 443 boxes were still on the truck. How many boxes did the truck start with?

b. Andy had $58.35. Then he spent some money. He ended up with $16.71. How much money did he spend?

c. A truck started out with 48 boxes. The truck picked up some more boxes. Now the truck has 94 boxes. How many boxes did the truck pick up?

Lesson 120

Part 5

5a.	b.	c.

a. Write 651 cents with a $ sign.

b. Write 380 cents with a $ sign.

c. Write 108 cents with a $ sign.

Part 6 Find the rows or the squares.

a.

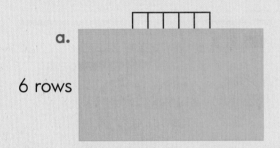

6 rows

b.

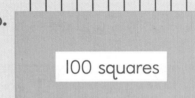

100 squares

6a.	
b.	

Part 7

a. How many cents is 12 nickels?

b. How many cents is 12 dimes?

c. How many cents is 4 quarters?

7a.	
b.	
c.	

Part 8 Copy and work each problem.

8a.	b.
c.	d.

a. $120 - 10 = \blacksquare$ b. $70 - 40 = \blacksquare$

c. $80 - 50 = \blacksquare$ d. $160 - 50 = \blacksquare$

Lesson 121

Part 1

a.
PM   PM

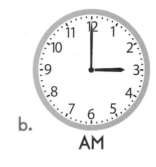
b.
AM   PM

c.
PM   AM

1a.	b.
c.	

Part 2

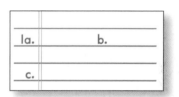

	0	1	2	3	4	5	6	7	8	9	10	11	12
Yard A													
Yard B													
Yard C													
Yard D													
Yard E													

Trees

1. Which yard has 5 trees?

2. Which yard has the most trees?

3. Which yard has the fewest trees?

4. Which yard has more trees than yard E?

5. How many trees does yard A have?

6. How many trees does yard C have?

2 1.	
2.	
3.	
4.	
5.	
6.	

Lesson 121

Part 3

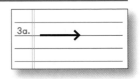

a.

Independent Work

Part 4 Copy and work each problem.

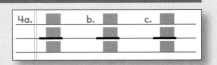

a. $12.40   b. $4.68   c. $19.37
   + 8.89   +9.38   − 8.77

Part 5 Figure out how much Tina ends up with.

$22.00 $31.30 $6.88 $12.70 $23.06

Tina has $54.75

a. She buys the football and the book.

b. She buys the book and the gloves.

c. She buys the football, the glasses, and the pens.

Part 6 Figure out the missing number.

a. 47 ■→ 59   b. ■ 26→ 80   c. 13 109→ ■

Lesson 121

Part 7 Copy and work each problem.

a. $40 - 20 = \blacksquare$ b. $80 - 40 = \blacksquare$

c. $140 - 70 = \blacksquare$ d. $130 - 30 = \blacksquare$

7a.	b.
c.	d.

Part 8 Write the time for each clock.

8a.	b.	c.

a.

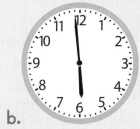

b.

c.

Part 9 Work each problem.

a. Mr. Green was 45 years old. Susan was 18 years old. How many years older was Mr. Green than Susan?

b. The piece of red wood was 25 inches longer than the piece of white wood. The piece of white wood was 152 inches long. How many inches long was the piece of red wood?

c. Sam weighed 17 pounds less than his father. Sam weighed 163 pounds. How many pounds did his father weigh?

9a. ⟶

b. ⟶

c. ⟶

Part 10 Find the perimeter and the area of each rectangle.

10a. $\blacksquare + \blacksquare = \blacksquare$ b. $\blacksquare + \blacksquare = \blacksquare$
$\blacksquare + \blacksquare = \blacksquare$ $\blacksquare + \blacksquare = \blacksquare$
$\blacksquare$ $\blacksquare$

$\blacksquare \times \blacksquare = \blacksquare$ $\blacksquare \times \blacksquare = \blacksquare$

10 in.

4 in.

a.

12 cm

2 cm

b.

Part 11 a. Write 136 cents with a $ sign.

b. Write 501 cents with a $ sign.

11a.	b.

Lesson 122

Part 1

1a.	b.	
c.	d.	

a.
PM   AM

b.
PM   PM

c.
AM   PM

d.
PM   AM

Part 2

a.

2a.	→
	→
	→
	→
	→

→

Independent Work

Part 3 Figure out the missing number.

3a.	b.
c.	

a. $5 \times 10 = \blacksquare$ b. $2 \times \blacksquare = 16$

c. $10 \times \blacksquare = 90$

Lesson 122

Part 4 Find the rows or the squares.

a.

8 rows

b.

45 squares

4a.	
b.	

Part 5 Work each problem.

a. Tom had 88 baseball cards. James had 136 baseball cards. Fran had 102 cards. How many cards did the children have altogether?

b. There were 234 children in the park. 106 were girls. How many were boys?

c. Nola had 24 books. Andy had 9 books. Dolly had 32 books. How many books did the children have altogether?

5a.	
b.	
c.	

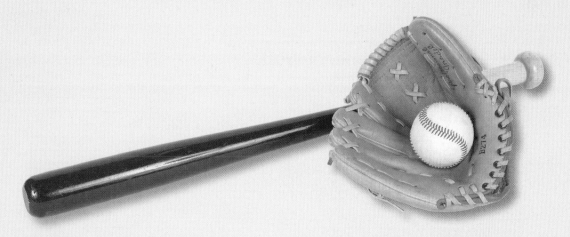

Part 6 Copy and work each problem.

a. 278
   +654

b. 792
   −167

c. 813
   −173

Connecting Math Concepts

Lesson 123

Part 1

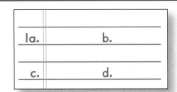

la.	b.
c.	d.

a.

 PM    AM

b.

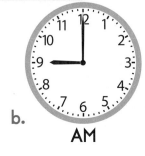 AM    PM

c.

 AM    AM

d.

 AM    PM

Part 2

a. 7

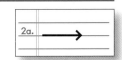

Independent Work

Part 3   Work each problem.

3a. ⟶

a. The red boat was 27 feet shorter than the
white boat. The white boat was 65 feet long.
How many feet long was the red boat?

b. ⟶

b. The big tree was 253 feet tall. The small tree
was 37 feet tall. How many feet taller was the
big tree than the small tree?

c. ⟶

c. There were 678 fish in a lake last year. There
are 896 fish in the lake this year. How many
more fish are in the lake this year?

Connecting Math Concepts

Lesson 123

Part 4
  a. Write 500 cents with a $ sign.

  b. Write 320 cents with a $ sign.

  c. Write 303 cents with a $ sign.

4a.	b.	c.

Part 5 Figure out the missing number.

a. 139  → 193

b. ■ 52 → 90

5a.	b.

Part 6 Write the time for each clock.

a.   b.

6a.	b.

Part 7 Work each problem.

  a. How many cents is 15 nickels?

  b. How many cents is 10 quarters?

  c. How many cents is 15 dimes?

7a.
b.
c.

Part 8

	red	white	blue	pink
Jean	2	15	10	8
Ann	12	3	5	6
Bob	9	6	11	3

Flowers

8a.
b.
c.
d.

a. Who has the most blue flowers?

b. Who has the fewest white flowers?

c. How many pink flowers does Jean have?

d. Who has 9 red flowers?

Lesson 124

a.  13

$$\underrightarrow{12 \quad 1} 13$$

$$\underrightarrow{11 \quad 2} 13$$

$$\underrightarrow{10 \quad 3} 13$$

$$\underrightarrow{9 \quad 4} 13$$

$$\underrightarrow{7 \quad 6} 13$$

$$\underrightarrow{6 \quad 7} 13$$

$$\underrightarrow{5 \quad 8} 13$$

$$\underrightarrow{4 \quad 9} 13$$

$$\underrightarrow{3 \quad 10} 13$$

$$\underrightarrow{2 \quad 11} 13$$

$$\underrightarrow{1 \quad 12} 13$$

b.  11

$$\underrightarrow{10 \quad 1} 11$$

$$\underrightarrow{9 \quad 2} 11$$

$$\underrightarrow{8 \quad 3} 11$$

$$\underrightarrow{7 \quad 4} 11$$

$$\underrightarrow{6 \quad 5} 11$$

$$\underrightarrow{4 \quad 7} 11$$

$$\underrightarrow{3 \quad 8} 11$$

$$\underrightarrow{2 \quad 9} 11$$

$$\underrightarrow{1 \quad 10} 11$$

c. 9

$$\underrightarrow{7 \quad 2} 9$$

$$\underrightarrow{6 \quad 3} 9$$

$$\underrightarrow{5 \quad 4} 9$$

$$\underrightarrow{4 \quad 5} 9$$

$$\underrightarrow{3 \quad 6} 9$$

$$\underrightarrow{2 \quad 7} 9$$

Lesson 124

Part 2

	0	1	2	3	4	5	6	7	8	9	10	11	12
yard A													
yard B													
yard C													
yard D													
yard E													

Trees

1. How many birds are in yard A?

2. How many birds are in yard C?

3. How many birds are in yard E?

4. Which yard has the most birds?

5. Which yard has the fewest birds?

6. How many birds are in the yard with the fewest birds?

7. Write the letters of the yards that have more than 2 birds.

2 1.	
2.	
3.	
4.	
5.	
6.	
7.	

Part 3 For each item, figure out how many hours go by.

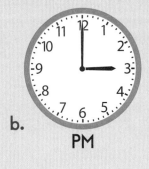

a.
AM   PM   b.   PM   AM

c.
PM   AM

3a.	b.
c.	

Lesson 124

Part 4 Find the perimeter and the area of each rectangle.

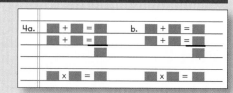

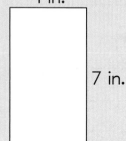

4 in.

7 in.

a.

10 cm

11 cm

b.

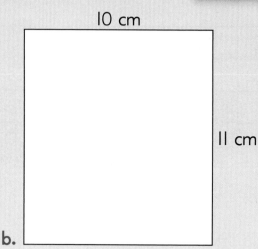

Part 5 Work each problem.

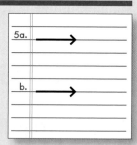

a. There were 87 birds in trees and 145 birds on the ground. How many birds were there in all?

b. There were 166 pencils in a store. 48 of the pencils were red. How many were not red?

Part 6

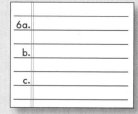

a. How many cents is 6 dimes?

b. How many cents is 6 nickels?

c. How many cents is 6 quarters?

Part 7 Figure out the missing number.

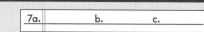

a. $4 \times \blacksquare = 20$ b. $4 \times \blacksquare = 36$ c. $4 \times 7 = \blacksquare$

Lesson 125

Part 1

a. About how many feet is 2 meters?

b. About how many feet is 4 meters?

1a.	
b.	

Part 2

a. 14

$$\xrightarrow[\quad 13 \quad 1 \quad]{} 14$$

$$\xrightarrow[\quad 12 \quad 2 \quad]{} 14$$

$$\xrightarrow[\quad 10 \quad 4 \quad]{} 14$$

$$\xrightarrow[\quad 9 \quad 5 \quad]{} 14$$

$$\xrightarrow[\quad 8 \quad 6 \quad]{} 14$$

$$\xrightarrow[\quad 6 \quad 8 \quad]{} 14$$

$$\xrightarrow[\quad 5 \quad 9 \quad]{} 14$$

$$\xrightarrow[\quad 4 \quad 10 \quad]{} 14$$

$$\xrightarrow[\quad 3 \quad 11 \quad]{} 14$$

$$\xrightarrow[\quad 2 \quad 12 \quad]{} 14$$

$$\xrightarrow[\quad 1 \quad 13 \quad]{} 14$$

b. 12

$$\xrightarrow[\quad 11 \quad 1 \quad]{} 12$$

$$\xrightarrow[\quad 10 \quad 2 \quad]{} 12$$

$$\xrightarrow[\quad 9 \quad 3 \quad]{} 12$$

$$\xrightarrow[\quad 8 \quad 4 \quad]{} 12$$

$$\xrightarrow[\quad 7 \quad 5 \quad]{} 12$$

$$\xrightarrow[\quad 6 \quad 6 \quad]{} 12$$

$$\xrightarrow[\quad 5 \quad 7 \quad]{} 12$$

$$\xrightarrow[\quad 4 \quad 8 \quad]{} 12$$

$$\xrightarrow[\quad 3 \quad 9 \quad]{} 12$$

$$\xrightarrow[\quad 2 \quad 10 \quad]{} 12$$

$$\xrightarrow[\quad 1 \quad 11 \quad]{} 12$$

c. 10

$$\xrightarrow[\quad 9 \quad 1 \quad]{} 10$$

$$\xrightarrow[\quad 8 \quad 2 \quad]{} 10$$

$$\xrightarrow[\quad 7 \quad 3 \quad]{} 10$$

$$\xrightarrow[\quad 6 \quad 4 \quad]{} 10$$

$$\xrightarrow[\quad 5 \quad 5 \quad]{} 10$$

$$\xrightarrow[\quad 4 \quad 6 \quad]{} 10$$

$$\xrightarrow[\quad 3 \quad 7 \quad]{} 10$$

$$\xrightarrow[\quad 1 \quad 9 \quad]{} 10$$

d. 11

$$\xrightarrow[\quad 9 \quad 2 \quad]{} 11$$

$$\xrightarrow[\quad 8 \quad 3 \quad]{} 11$$

$$\xrightarrow[\quad 7 \quad 4 \quad]{} 11$$

$$\xrightarrow[\quad 6 \quad 5 \quad]{} 11$$

$$\xrightarrow[\quad 4 \quad 7 \quad]{} 11$$

$$\xrightarrow[\quad 3 \quad 8 \quad]{} 11$$

$$\xrightarrow[\quad 2 \quad 9 \quad]{} 11$$

$$\xrightarrow[\quad 1 \quad 10 \quad]{} 11$$

Lesson 125

Part 3

a. Jane was 7 years older than Dan. Dan was 23 years old. How many years old was Jane?

b. There were blue trucks and red trucks on the street. 13 trucks were blue. There were 30 trucks in all. How many red trucks were there?

c. A truck started out with 50 boxes. Then workers took 19 boxes from the truck. How many boxes were still on the truck?

d. 17 boys and 26 girls were in the park. How many children were in the park?

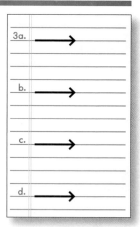

Independent Work

Part 4 Find the rows or the squares.

a.

6 rows

b.

100 squares

Part 5

Each subtraction problem shows how much money each person started with and how much each person spent.

Harry	$250 − $80
Debby	$250 − $140
Tom	$250 − $90

a. Which person spent the most?

b. How much did that person spend?

c. Which person ended up with the most?

d. How much did that person end up with?

Lesson 125

Part 6 For each item, figure out how many hours go by.

6a.	b.
c.	d.

a. AM   AM

b. PM   AM

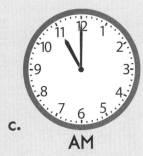

c. AM   PM

d. AM   PM

Part 7 Work the estimation problem for each item.

7a.	b.
c.	d.

a. $56 + 12 = \blacksquare$ b. $18 + 89 = \blacksquare$

c. $77 - 23 = \blacksquare$ d. $41 + 98 = \blacksquare$

Part 8 Work the column problem to figure out each missing number.

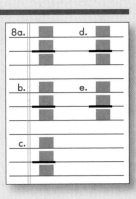

a. $56 - \blacksquare = 33$ d. $\blacksquare - 14 = 89$

b. $\blacksquare - 80 = 200$ e. $\blacksquare + 280 = 285$

c. $123 + \blacksquare = 444$

Lesson 126

Part 1

a. About how many feet is 2 meters?

b. About how many feet is 5 meters?

c. About how many feet is 3 meters?

1a.	
b.	
c.	

Part 2

a. The boat was 55 feet long. The truck was 37 feet long. How many feet longer was the boat than the truck?

b. There were 45 glasses on a table. Henry put 12 more glasses on the table. How many glasses ended up on the table?

c. There were 16 fewer cars than trucks on the street. There were 26 cars on the street. How many trucks were there?

d. There were elm trees and pine trees in a park. There were 90 trees in all. 58 were pine trees. How many elm trees were there?

2a.	→
b.	→
c.	→
d.	→

Independent Work

Part 3

Make number families to show all the ways you can make two groups of balls.

a.  $\overset{4}{\underset{\longrightarrow}{\quad}}\overset{1}{}5$

3a.	→

Part 4

a. How many cents is 9 nickels?

b. How many cents is 9 dimes?

c. How many cents is 5 quarters?

4a.	
b.	
c.	

Connecting Math Concepts

Lesson 126

Part 5 For each problem, figure out the number of rows or the number of squares.

a.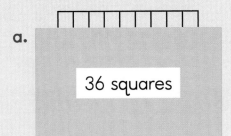

36 squares

b.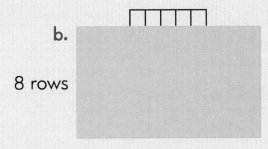

8 rows

c.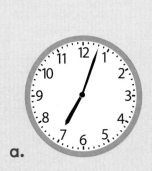

16 squares

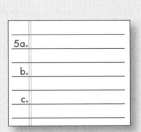

5a.

b.

c.

Part 6 Write the time for each clock.

6a.   b.   c.

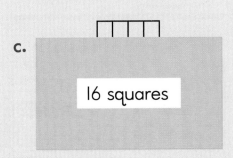

a.

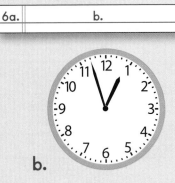

b.

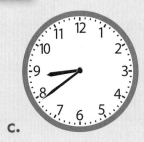

c.

Lesson 126

Part 7 Find the perimeter and the area of each rectangle.

5 ft

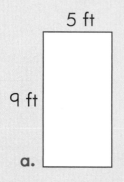

9 ft

a.

4 m

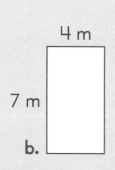

7 m

b.

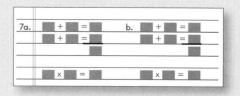

Part 8 Work the estimation problem for each item.

a. $51 - 39 = $ ■

b. $28 + 59 = $ ■

c. $69 - 22 = $ ■

d. $48 + 49 = $ ■

e. $42 - 18 = $ ■

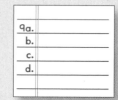

Part 9 Answer each question.

Pots in Different Kitchens

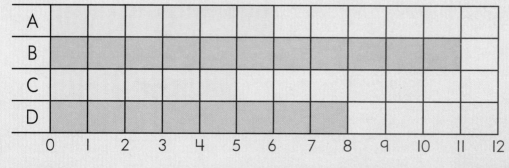

a. Which kitchen has the most pots?

b. How many kitchens have more than 5 pots?

c. Which kitchen has the fewest pots?

d. How many pots are in kitchen A?

Lesson 127

Part 1

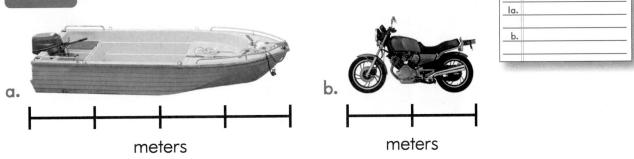

a.

meters

b.

meters

1a.	
b.	

Part 2

a. The green house was 82 years older than the white house. The white house was 104 years old. How many years old was the green house?

b. There were short books and long books on the shelf. There were 46 books. 18 of the books were short. How many long books were on the shelf?

c. The shelf had books on it. Then a woman took 78 books from the shelf. The shelf ended up with 20 books on it. How many books did the shelf start out with?

d. Mr. Briggs was 89 years old. Mr. Green was 46 years old. How many years older was Mr. Briggs than Mr. Green?

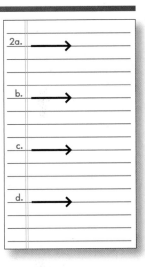

2a.	→
b.	→
c.	→
d.	→

Lesson 127

Part 3 Make number families to show all the ways you can make two groups of balls.

a.

Part 4 Answer each question.

Color of Houses

	blue	white	green
A Street	7	10	3
B Street	9	6	15
C Street	4	10	7
D Street	6	12	8

a. Which street had 7 green houses?

b. Which street had the most white houses?

c. Were there more blue houses, white houses, or green houses on D Street?

d. Were there more blue houses, white houses, or green houses on B Street?

e. Which 2 streets had the same number of white houses?

Part 5 Copy and work each problem.

a. $120 - 10 = \blacksquare$ b. $70 - 40 = \blacksquare$

c. $80 - 50 = \blacksquare$ d. $160 - 80 = \blacksquare$

Lesson 127

Part 6

Each subtraction problem shows how much money each person started with and how much each person spent.

Donna	$87 − $12
Margie	$87 − $85
Ann	$87 − $9

6a. _____
b. _____
c. _____
d. _____

a. Which person spent the most?

b. How much did that person spend?

c. Which person ended up with the most?

d. How much did that person end up with?

Part 7 For each item, figure out how many hours go by.

7a. _____ b. _____
c. _____ d. _____

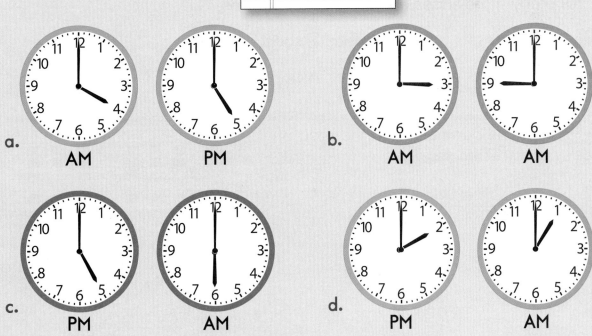

a. AM PM

b. AM AM

c. PM AM

d. PM AM

Part 8 Work the column problem to figure out each missing number.

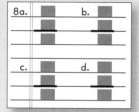

a. $\blacksquare - 24 = 155$ b. $56 + \blacksquare = 179$

c. $\blacksquare + 120 = 338$ d. $56 + \blacksquare = 88$

Part 9 Answer each question.

Shoes in Different Closets

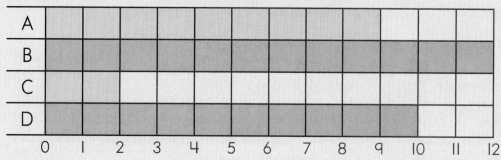

A												
B												
C												
D												
0	1	2	3	4	5	6	7	8	9	10	11	12

a. Which closet has the most shoes?

b. Which closet has the fewest shoes?

c. How many more shoes are in closet B than closet D?

d. How many closets have more than 7 shoes?

e. What is the total number of shoes in closet A and closet C?

Part 10 Find the perimeter and the area of each rectangle.

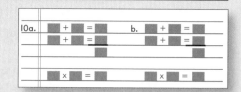

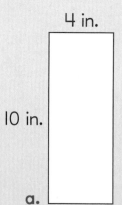

4 in.

10 in.

a.

9 cm

b.

2 cm

Lesson 128

Part 1

a. Karen started out with 41 cards. Then she found some more cards. She ended up with 70 cards. How many cards did she find?

b. In a garden, there were 60 pink roses and 45 white roses. How many more pink roses than white roses were in the garden?

c. There were red flowers and yellow flowers in the garden. 47 flowers were red. 55 flowers were yellow. How many flowers were in the garden?

d. The white dog weighed 66 fewer pounds than the brown dog. The white dog weighed 18 pounds. How many pounds did the brown dog weigh?

1a.	→
b.	→
c.	→
d.	→

Independent Work

Part 2 Answer each question.

a. About how many feet is 5 meters?

b. About how many feet is 4 meters?

2a.	
b.	

Part 3 Work the estimation problem for each item.

a. $57 - 13 = \blacksquare$ d. $11 + 39 = \blacksquare$

b. $68 - 22 = \blacksquare$ e. $52 + 57 = \blacksquare$

c. $31 - 19 = \blacksquare$

3a.	d.
b.	e.
c.	

Lesson 128

Part 4 Work each item.

$27.30 $49.00 $64.99 $31.50

$169.75

4a.

b.

a. If Juan buys the radio, the shoes, and the book, how much will he still have?

b. If Juan buys the radio and the book, how much will he still have?

Part 5 Work the column problem to figure out each missing number.

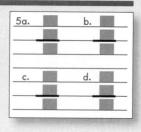

a. $120 + \blacksquare = 326$ b. $\blacksquare - 135 = 328$

c. $490 - \blacksquare = 179$ d. $\blacksquare - 59 = 289$

Part 6 Find the perimeter and the area of each rectangle.

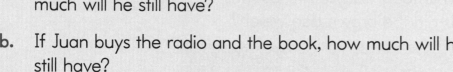

9 in.

a.

9 in.

b.

7 yd

5 yd

Lesson 129

Part 1

a. There were 17 mice in the field and 66 mice in the barn. How many fewer mice were in the field than the barn?

b. The train started out with 28 people on it. At the first stop, more people got on. The train ended up with 241 people on it. How many people got on at the first stop?

c. There were 19 yellow cups and 38 green cups on the shelf. How many cups were on the shelf?

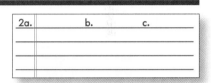

Part 2 Find the perimeter of each figure.

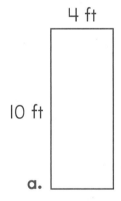

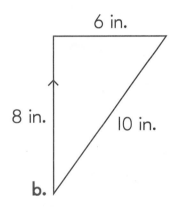

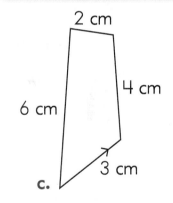

Level C Correlation to Grade 2
Common Core State Standards for Mathematics

Operations and Algebraic Thinking (2.OA)

Represent and solve problems involving addition and subtraction.

1. Use addition and subtraction within 100 to solve one- and two-step word problems involving situations of adding to, taking from, putting together, taking apart, and comparing, with unknowns in all positions, e.g., by using drawings and equations with a symbol for the unknown number to represent the problem.

Lessons	WB 1: 33–41, 65
	WB 2: 93, 94, 107, 121–125, 127, 128, 130
	TB: 42–100, 102, 103, 105–112, 114–116, 118–120, 123, 129

Operations and Algebraic Thinking (2.OA)

Add and subtract within 20.

2. Fluently add and subtract within 20 using mental strategies. By end of Grade 2, know from memory all sums of two one-digit numbers.

Lessons	WB 1: 1–39, 41–70
	WB 2: 71–80, 82, 85–122, 129, 130
	TB:42–49, 56, 58, 59, 62–64, 68–73, 75, 77, 78, 81–83, 85–87, 95, 96, 99–106, 110–115, 119, 121–127

Operations and Algebraic Thinking (2.OA)

Work with equal groups of objects to gain foundations for multiplication.

3. Determine whether a group of objects (up to 20) has an odd or even number of members, e.g., by pairing objects or counting them by 2s; write an equation to express an even number as a sum of two equal addends.

Lessons	WB 1: 55–60
	WB 2: 128, 129

Operations and Algebraic Thinking (2.OA)

Work with equal groups of objects to gain foundations for multiplication.

4. Use addition to find the total number of objects arranged in rectangular arrays with up to 5 rows and up to 5 columns; write an equation to express the total as a sum of equal addends.

Lessons	WB 1: 47, 48
	WB 2: 71, 73
	TB: 45–47, 49, 75

Number and Operations in Base Ten (2.NBT)

Understand place value.

1. Understand that the three digits of a three-digit number represent amounts of hundreds, tens, and ones; e.g., 706 equals 7 hundreds, 0 tens, and 6 ones. Understand the following as special cases:
   a. 100 can be thought of as a bundle of ten tens — called a "hundred."
   b. The numbers 100, 200, 300, 400, 500, 600, 700, 800, 900 refer to one, two, three, four, five, six, seven, eight, or nine hundreds (and 0 tens and 0 ones).

Lessons	WB 1: 1–4, 7, 10–25, 27, 29, 36, 59, 68
	WB 2: 74, 116, 118, 126–129
	Student Practice Software: Block 1 Activities 1 and 2, Block 4 Activities 2 and 3, Block 6 Activity 6

Number and Operations in Base Ten (2.NBT)

Understand place value.

2. Count within 1000; skip-count by *2, 5s, 10s, and 100s.

Lessons	WB 1: 11, 12, 15–29, 31–60, 63–65, 67–69
	WB 2: 73–77, 80, 88, 90–94, 98, 101, 102, 116–120, 122–126, 130
	TB: 41, 62, 63, 65, 68, 69, 72, 73, 76–80, 86, 87, 89 ,90, 95–117, 119, 120, 122, 125–127
	Student Practice Software: Block 3 Activity 2

*Denotes California-only content.

Number and Operations in Base Ten (2.NBT)

Understand place value.

3. Read and write numbers to 1000 using base-ten numerals, number names, and expanded form.

Lessons	WB 1: 1–7, 11–37, 39, 40, 42, 43, 45–51, 54, 58, 59, 68
	WB 2: 74
	TB: 82, 118
	Student Practice Software: Block 2 Activity 3, Block 4 Activity 4, Block 5 Activity 4

Number and Operations in Base Ten (2.NBT)

Understand place value.

4. Compare two three-digit numbers based on meanings of the hundreds, tens, and ones digits, using >, =, and < symbols to record the results of comparisons.

Lessons	WB 1: 31, 36
	WB 2: 112, 113, 115, 119–121, 124–128
	Student Practice Software: Block 2 Activity 5

Number and Operations in Base Ten (2.NBT)

Use place value understanding and properties of operations to add and subtract.

5. Fluently add and subtract within 100 using strategies based on place value, properties of operations, and/or the relationship between addition and subtraction.

Lessons	WB 1: 1–70 WB 2: 71–91, 93–130 TB: 47–50, 52–91, 94–96, 98–100, 102–115, 117–129 Student Practice Software: Block 3 Activities 1, 3, 4, 5; Block 5 Activity 2

Number and Operations in Base Ten (2.NBT)

Use place value understanding and properties of operations to add and subtract.

6. Add up to four two-digit numbers using strategies based on place value and properties of operations.

Lessons	WB 1: 4–10, 12, 16–20, 22–56, 60–70 WB 2: 71–94, 96–98, 100–102, 110, 111, 113, 122, 127–130 TB: 46, 47, 49, 50, 52–55, 57, 59–66, 69–71, 73, 75–82, 84, 86, 88, 89, 92–100, 102, 107–114, 117, 119, 122, 124, 125, 126, 128 Student Practice Software: Block 1 Activity 3

Number and Operations in Base Ten (2.NBT)

Use place value understanding and properties of operations to add and subtract.

7. Add and subtract within 1000, using concrete models or drawings and strategies based on place value, properties of operations, and/or the relationship between addition and subtraction; relate the strategy to a written method. Understand that in adding or subtracting three-digit numbers, one adds or subtracts hundreds and hundreds, tens and tens, ones and ones; and sometimes it is necessary to compose or decompose tens or hundreds.

Lessons	WB 1: 5–8, 10–70 WB 2: 71, 73–89, 92–95, 100, 106, 107, 109, 111–115, 119–121, 124, 125, 127–130 TB: 46–50, 62–69, 71, 73–83, 85, 86, 88–117, 119–123, 125, 127, 128 Student Practice Software: Block 1 Activity 5, Block 2 Activity 2, Block 5 Activity 5

Number and Operations in Base Ten (2.NBT)

Use place value understanding and properties of operations to add and subtract.

*7.1 Use estimation strategies to make reasonable estimates in problem solving.

Lessons	WB 2: 75–84, 88, 90–94, 97, 98, 100, 102 TB: 76–79, 95, 96, 98, 99, 105, 107–109, 111–114, 125, 126, 128

*Denotes California-only content.

Number and Operations in Base Ten (2.NBT)

Use place value understanding and properties of operations to add and subtract.

8. Mentally add 10 or 100 to a given number 100–900, and mentally subtract 10 or 100 from a given number 100–900.

Lessons	WB 1: 22–24, 42, 43, 50, 58, 59 WB 2: 75–78, 80, 89, 112, 117–121 TB: 89

Number and Operations in Base Ten (2.NBT)

Use place value understanding and properties of operations to add and subtract.

9. Explain why addition and subtraction strategies work, using place value and the properties of operations.

Lessons	WB 2: 115–120

Measurement and Data (2.MD)

Measure and estimate lengths in standard units.

1. Measure the length of an object by selecting and using appropriate tools such as rulers, yardsticks, meter sticks, and measuring tapes.

Lessons	WB 1: 30–41, 43, 44, 46, 49, 54, 59–68 WB 2: 115–120, 122, 124–127

Measurement and Data (2.MD)

Measure and estimate lengths in standard units.

2. Measure the length of an object twice, using length units of different lengths for the two measurements; describe how the two measurements relate to the size of the unit chosen.

Lessons	WB 1: 34, 35, 41, 43, 44, 46, 49, 51, 53, 54 WB 2: 85–87, 116

Measurement and Data (2.MD)

Measure and estimate lengths in standard units.

3. Estimate lengths using units of inches, feet, centimeters, and meters.

Lessons	WB 2: 85–87 TB: 125–127

Measurement and Data (2.MD)

Measure and estimate lengths in standard units.

4. Measure to determine how much longer one object is than another, expressing the length difference in terms of a standard length unit.

Lessons	WB 2: 115–120, 122

Measurement and Data (2.MD)

Relate addition and subtraction to length.

5. Use addition and subtraction within 100 to solve word problems involving lengths that are given in the same units, e.g., by using drawings (such as drawings of rulers) and equations with a symbol for the unknown number to represent the problem.

Lessons	TB: 52, 53, 57–71, 74, 75, 83, 84, 94, 96, 98, 123, 126

Measurement and Data (2.MD)

Relate addition and subtraction to length.

6. Represent whole numbers as lengths from 0 on a number line diagram with equally spaced points corresponding to the numbers 0, 1, 2, ..., and represent whole-number sums and differences within 100 on a number line diagram.

Lessons	WB 1: 51–54, 56, 59, 67–70 WB 2: 125, 127, 128 Student Practice Software: Block 6 Activity 1

Measurement and Data (2.MD)

Work with time and money.

7. Tell and write time from analog and digital clocks to the nearest five minutes, using a.m. and p.m. ***Know relationships of time (e.g., minutes in an hour, days in a month, weeks in a year).**

Lessons	WB 2: 91–107, 109, 112, 114, 116, 118, 120, 122–124, 130 TB: 85–88, 90, 98, 99, 101, 103, 106, 108, 109, 112, 115, 117–119, 121–123, 126 Student Practice Software: Block 6 Activity 2

*Denotes California-only content.

Measurement and Data (2.MD)

Work with time and money

8. Solve word problems involving dollar bills, quarters, dimes, nickels, and pennies, using $ and ¢ symbols appropriately. *Example: If you have 2 dimes and 3 pennies, how many cents do you have?*

Lessons	WB 2: 94, 130 TB: 43, 46, 48, 54–56, 58, 62, 64, 65, 73, 75, 76, 80, 86, 87, 90, 93, 97–116, 118–121, 123, 124, 126, 128 Student Practice Software: Block 4 Activity 3, Block 5 Activity 6

Measurement and Data (2.MD)

Represent and interpret data.

9. Generate measurement data by measuring lengths of several objects to the nearest whole unit, or by making repeated measurements of the same object. Show the measurements by making a line plot, where the horizontal scale is marked off in whole-number units.

Lessons	WB 2: 124–127 Student Practice Software: Block 6 Activity 3

Measurement and Data (2.MD)

Represent and interpret data.

10. Draw a picture graph and a bar graph (with single-unit scale) to represent a data set with up to four categories. Solve simple put-together, take-apart, and compare problems using information presented in a bar graph.

Lessons	WB 2: 117–120, 122, 123, 125 TB: 121, 124, 126, 127 Student Practice Software: Block 6 Activity 4

Geometry (2.G)

Reason with shapes and their attributes.

1. Recognize and draw shapes having specified attributes, such as a given number of angles or a given number of equal faces. Identify triangles, quadrilaterals, pentagons, hexagons, and cubes.

Lessons	WB 1: 43–49, 54, 56–58 WB 2: 111–118, 127–129 TB: 50, 51, 53, 55–57, 59, 60, 66, 68, 70, 74, 76, 79, 84, 110 Student Practice Software: Block 4 Activity 5

Geometry (2.G)

Reason with shapes and their attributes.

2. Partition a rectangle into rows and columns of same-size squares and count to find the total number of them.

Lessons	WB 1: 47, 48 WB 2: 121–126, 128 TB: 45, 46, 48, 49, 126 Student Practice Software: Block 4 Activity 6

Geometry (2.G)

Reason with shapes and their attributes.

3. Partition circles and rectangles into two, three, or four equal shares, describe the shares using the words *halves, thirds, half of, a third of,* etc., and describe the whole as two halves, three thirds, four fourths. Recognize that equal shares of identical wholes need not have the same shape.

Lessons	WB 2: 125–127 Student Practice Software: Block 6 Activity 5

Standards for Mathematical Practice

Connecting Math Concepts addresses all of the Standards for Mathematical Practice throughout the program. What follows are examples of how individual standards are addressed in this level.

1. Make sense of problems and persevere in solving them.

Word Problems (Lessons 12–25, 28–77): Students learn to identify specific types of word problems (i.e., start-end, comparison, classification) and set up and solve the problems based on the specific problem types.

2. Reason abstractly and quantitatively.

Addition/Subtraction (Lessons 1–64, 73–115): Beginning in Lesson 45, students count objects in two groups, write the number for each group, and then add them to find the total objects. They connect the written numbers with quantities while learning the concept of addition.

3. Construct viable arguments and critique the reasoning of others.

Estimation (Lessons 65–94): Students learn how to round numbers and then apply that knowledge to word problems involving estimation. They work the original problem and the estimation problem and then compare answers to verify that the estimated answer is close to the exact answer. Students can construct an argument to persuade someone whether an estimated answer is reasonable.

4. Model with mathematics.

Number Families (Lessons 1–29): Students learn to represent three related numbers in a number family. Later, they apply number families to model and solve word problems.

5. Use appropriate tools strategically.

Throughout the program (Lessons 1–130) students use pencils, workbooks, lined paper, and textbooks to complete their work. They use rulers to measure lines. They use the computer to access the Practice Software where they apply the skills they learn in the lessons.

6. Attend to precision.

Measurement (Lessons 30–43, 59–78, 95–103): When measuring lines and finding perimeter and area, students learn to include the correct unit in the verbal and written answers. They also include units in answers to word problems that involve specific units.

7. Look for and make use of structure.

Mental Math (Lesson 9 and frequently throughout): Students build computational fluency by learning patterns inherent in different problem types, such as +/– 10 and +/– 100.

8. Look for and express regularity in repeated reasoning.

Multiplication (Lessons 32–49): Students are introduced to the concept of multiplication by thinking of it as repeated counting. For example, 2 x 5 tells us to count by 2 five times.

Photo Credits